THE
ELECTRONICS REPAIR COOKBOOK

THE ELECTRONICS REPAIR COOKBOOK

Save Time and Money by Fixing Electronics Yourself

MARK RUMREICH

Library of Congress Control Number: 2020920430

Copyright © 2021 by Mark Rumreich. All rights reserved. Except as permitted under the United States Copyright Act of 1976, no part of this publication may be reproduced or distributed in any form or by any means, or stored in a data base or retrieval system, without the prior written permission of the author.

ISBN 979-8-6863-4381-8

Cover design by Matchbook.

Information contained in this work has been obtained from sources believed to be reliable. However, the author does not guarantee the accuracy or completeness of any information published herein and the author shall not be responsible for any errors, omissions, or damages arising out of use of this information. This work is published with the understanding that the author is supplying information but is not attempting to render engineering or other professional services. If such services are required, the assistance of an appropriate professional should be sought.

Printed in the United States.

First printing.
10 9 8 7 6 5 4 3 2 1

To Grandpa Frank

Contents

Acknowledgments — xvii

CHAPTER 1
Before You Begin   1

Introduction — 1
How This Book Is Organized — 2

CHAPTER 2
Safety   5

High Voltage — 5
    Electrolytic Capacitors — 5
    Inverter Circuits — 6
Lasers — 7
Solder and Soldering — 7
Solvents and Chemicals — 7
Eye Protection — 8
Childproof Your Work Area — 8
Don't Work Alone — 8
Electronic Waste Disposal — 8

CHAPTER 3
Electronics Basics 9

Introduction to Components	9
Resistors	9
Potentiometers	9
Capacitors	11
Inductors	11
Diodes	12
Transistors	12
Integrated Circuits	14
Transformers	14
Relays	15
Introduction to Schematics	16
Block Diagrams versus Schematics	16
Component Symbols	18
Introduction to Electricity	19
Basic Units of Electricity	19
Ohm's Law	20
Power Formulas	20
Batteries in Series	20
Resistors in Series and Parallel	21
Capacitors in Parallel	22
AC versus DC	22

CHAPTER 4
Equipment and Supplies 25

Basic Supplies	25
Electrical Contact Cleaner Sprays	25
Air Duster	26
Freeze Mist	27
Solder	28
Solder Wick	28
Heat-Shrink Tubing	28

Electrical Tape	30
Heat-Sink Compound	30
Nail Polish	31
Wire Ties	31
Solvents	32
Glues	33
Clip Leads	35
Crimp Connectors	36
Hand Tools	38
Needle-Nose Pliers	38
Wire Cutters	38
Wire Stripper	39
X-Acto Knife	40
Precision Screwdriver Set	40
JIS (Japanese Industrial Standard) Screwdriver Set	42
Security-Bit Set	44
Solder Sucker	45
Tweezers	45
Hemostats	45
Pin Vise	46
Dental Pick	46
Crimper	46
Strain-Relief Pliers	47
Nut-Driver Set	47
Basic Equipment	48
Soldering Iron/Soldering Station	48
Digital Multimeter	50
Magnifiers: OptiVISOR/Magnifier Lamp	51
Antistatic Mat and Wrist Strap	53
Intermediate Equipment	54
Heat Gun	54
Digital Calipers	55
Panavise	56
Dummy Loads	56

Advanced Equipment ... 58
    Power Supply ... 59
    ESR Meter ... 60
    Oscilloscope ... 62
    Function Generator ... 64
    Isolation Transformer ... 64

CHAPTER 5

Soldering and Desoldering   67

Soldering ... 67
    Proper Tip Size and Temperature ... 67
    Use the Right Solder ... 68
    How to Solder ... 71
    Flux Removal after Soldering ... 72
Desoldering ... 72
    Solder Wick ... 73
    Solder Suckers ... 74
    Flux Pen ... 74
    Add Solder ... 74
    Use a Pin Vise ... 75
    Clip Instead of Desolder ... 76
PC Board Trace Repair ... 78
Heat-Sinked Devices ... 80
    Heat-Sink Compound ... 80
    Thermal Pads ... 81
    Mica Insulators ... 83
    Bolt Down, Then Solder ... 83
Soldering and Desoldering Surface-Mount Parts ... 84

CHAPTER 6

Using Test Equipment   87

Using a Digital Multimeter ... 87
    General Measurement Procedure ... 87
    Continuity Mode ... 90

Current Measurement	90
Diode Test	91
Using an Oscilloscope	93
Scope Probes	93
Probe Compensation	94

CHAPTER 7

Part Identification, Testing, and Substitution   97

Counterfeit and Off-Brand Parts	97
Resistors	98
Identifying Resistance Value and Tolerance	99
Power Rating	102
Substitution	102
Potentiometers	103
Pots Used as Controls	104
Pots Used for Calibration	104
Other Pot Parameters and Substitution	105
Capacitors	106
Electrolytic Capacitors	106
Discharging Electrolytic Capacitors	108
Replacing Electrolytic Capacitors	110
Tantalum Capacitors	111
Film Capacitors	113
Ceramic Capacitors	113
Replacing Ceramic Capacitors	115
Inductors	115
Replacing Inductors	116
Diodes	117
Signal Diodes	118
Rectifier Diodes	118
Schottky Diodes	118
Zener Diodes	119
Varactor Diodes	119

Transistors	119
Transistor Packages	120
Transistor Failure Modes	121
Transistor Substitution	123
Voltage Regulators	124
Checking Regulators	125
Integrated Circuits	125
IC Packages	126
IC Testing	126
IC Substitution	127
Varistors	128
Varistor Identification and Replacement	129
Thermistors	129
Thermistor Identification and Replacement	130
Fuses	130
Fuse Failure	131
Fuse Replacement	131
Switches	132
Pole and Throw Terminology	133
Switch Function	135
Try Contact Cleaner	135
Replacement and Substitution	135
Transformers	137
Transformer Failure	138
Transformer Replacement and Substitution	138
Relays	139
Relay Testing	140
Replacement and Substitution	140
Analog Meters	141
Stuck Needle Fixes	142
Meter Replacement and Substitution	143
AC Adapters	144
AC Adapter Basics	144
Testing AC Adapters for a Bad Cable	145

Testing AC Adapters for Correct Output Voltage 145
AC Adapter Repair 146
AC Adapter Substitution 149
Component Test Summary 150

CHAPTER 8
General Repairs and Troubleshooting 151
Is It Worth Fixing? 151
   Poor Candidates for Repair 152
How Did It Break? 152
First Things to Check 153
   Is It Getting Power? 153
   Check the Fuse 153
   Unplug for a Few Minutes 153
   Factory Reset 153
Taking It Apart 154
   Screws under Feet and Labels 154
   Hidden Tabs 155
   Cutting It Apart 156
   Keeping Track of Which Screws Go Where 158
   Take Pictures 160
   Board Connectors 160
Visual Inspection 164
   Bulging or Leaking Electrolytic Capacitors 164
   Signs of Liquid or Corrosion Damage 165
   Burned Components 165
   Deformed Components 165
   Melted Insulation on Wires 166
   Components with a Hairline Crack 166
   Bad Solder Joints 166
   Components with a Chunk Missing 167
   Discolored Areas on the PC Board or Cabinet 167
Likely Suspects 167
   Electrolytic and Tantalum Capacitors 167

High-Power Components	167
I/O Components	168
Mechanical Components	168
Intermittent Problems	169
Get the Problem to Show Up	169
Likely Suspects for Intermittent Problems	169
In-Circuit Component Testing	170
In-Circuit versus Out-of-Circuit Testing	170
Measure In-Circuit	171
Lift One Lead	172
The Joy of Sets	173
Power the Circuit	173
More Joy of Sets	176
Troubleshooting	176
What Is Troubleshooting?	177
Understand the System	177
Divide and Conquer	179
Fix the Cause	180
Power Supply Troubleshooting	180
Transistor Circuitry Troubleshooting	182
Keeping a Repair Log	187

CHAPTER 9

Product-Specific Repairs   191

Flat-Screen TVs	191
What's Inside a Flat-Screen TV?	192
Taking a Flat-Screen TV Apart	194
Diagnosing Flat-Screen TV Problems	196
Laptop Computers	202
Finding Replacement Laptop Parts	203
Laptop Cracked Screen Replacement	203
Laptop Keyboard Replacement	205
Rechargeable Battery Packs	207
Replace or Repair?	207

Identifying Cell Types	207
Rebuilding a Battery Pack	209
Audio Receivers and Power Amps	211
Electrical Contact Problems	211
Blown Output Amplifier Circuitry	212
Loudspeakers	217
Driver Damage	217
Driver Replacement	218
Listening Test and Fine-Tuning	219
Additional Considerations for Woofer Substitution	220
Woofer Surround Repair	221
To Shim or Not to Shim	222
Surround Size and Edge Type	224
Shimless Surround Repair	227
Remote Controls	228
User Error	228
Battery Problems	229
Keypad Problems	230
Replacement	231

Part Suppliers 233
References 235
Index 237

Acknowledgments

Thanks to Dennis Petruzzi and Kevin Williams for their thoughtful reviews of the manuscript. This book is better because of them.

Thank you to Donna Gray and her team at Matchbook for an inspired cover design.

Thanks to Judy Bass at Industrial Press for sharing the vision of another book and to Patty Wallenburg at TypeWriting for a fantastic page layout.

CHAPTER 1

Before You Begin

Introduction

Baking a cake shouldn't require you to be a professional baker. And fixing your TV or laptop shouldn't require you to have a background in electronics. That's the idea behind this book. The information in this book can save you time and hundreds of dollars with a single repair.

The explosion of do-it-yourself (DIY) repair videos and other online resources has fueled a surge of interest in electronics repair for beginners. The growing right-to-repair movement means that consumers will demand more products that they can fix themselves. But repair videos and online repair advice are not always accurate, and there are many gaps in the information. This book complements what's online and fills those gaps.

This book will give beginners the confidence and know-how to repair electronics. It starts from scratch with the basics and advances to more complex repair techniques. It's filled with photos, diagrams, and charts to make the material easy to understand. But even experienced repairers will benefit from the know-how in this book.

How This Book Is Organized

If you have no electronics background, start with Chapter 2 and work your way forward. If you do have some electronics experience, you'll be able to move quickly through the material in the early chapters. But don't skip them entirely; there are useful tips for experienced repairers in every chapter.

- **Chapter 2, "Safety."** This chapter discusses safe working practices for electronics repair. Some hazards are obvious, but others, such as charged capacitors in electronics that are unplugged and off or high voltage in battery-operated devices, are not so obvious.
- **Chapter 3, "Electronics Basics."** If you're new to electronics, this chapter will introduce you to the basics of electronic components, block and schematic diagrams, and electricity.
- **Chapter 4, "Equipment and Supplies."** Repairing electronics requires some specialized supplies and equipment. You may already have some of what you'll need. The real value of this chapter is that it lets you discover something to make repairs faster, easier, cheaper, or better. It's organized by basic supplies, hand tools, and equipment from basic to advanced.
- **Chapter 5, "Soldering and Desoldering."** Knowing how to solder and desolder is a necessary skill for electronics repair. With some knowledge and a little practice, everyone can learn to solder well enough for basic repair work. This chapter provides the knowledge; you'll have to do the practicing on your own. This chapter also covers printed circuit board (PCB) trace repair, replacing heat-sinked devices, and special techniques for surface-mount devices.
- **Chapter 6, "Using Test Equipment."** This chapter explains the basics of how to use the two most important pieces of test equipment for electronics repair: digital multimeters and oscilloscopes. This chapter shows what the equipment can be used for and its ease of use.
- **Chapter 7, "Part Identification, Testing, and Substitution."** This chapter is about how to identify parts, tell whether they're work-

ing, and find a substitute, if needed. The procedures for checking parts in this chapter assume that the part is out of circuit.
- **Chapter 8, "General Repairs and Troubleshooting."** This chapter provides repair guidance that can be applied to any electronic device. Topics include poor candidates for repair, taking it apart, what to look for in a visual inspection, likely suspects, intermittent problems, and in-circuit component testing. The second half of the chapter is devoted to the art of troubleshooting. Topics include resources for understanding the system, troubleshooting strategy, power supply troubleshooting, and transistor circuitry troubleshooting.
- **Chapter 9, "Product-Specific Repairs."** This chapter provides specific repair advice for the most commonly repaired products, including audio receivers, rechargeable battery packs, flat screen TVs, laptops, and remote controls.

CHAPTER 2

Safety

High Voltage

When repairing electronics, the most important safety concern is electrical shock. Any voltage higher than 50 volts has the potential to stop a human heart. If you're wet or the contact point penetrates the skin, even lower voltages can be lethal.

Any device that's powered by 120 volts AC and plugged in clearly has the potential to kill, but even devices that run on batteries or are unplugged can still electrocute you (see Table 2-1).

TABLE 2-1 Common Hazards in Electronics Repair

Hazard	Action
Device powered by 120-VAC line	Unplug from AC line while repairing
Stored charge in electrolytic capacitors	Discharge large capacitors
Inverter circuits	Remove power and batteries while repairing

Electrolytic Capacitors

Electrolytic capacitors can store high voltages for minutes or even hours. Large-capacity high-voltage electrolytics are common in the power supply circuits of many products and pose a potential hazard. You can use a DC voltmeter to check for electrolytics with a stored charge. It's a good idea to discharge any large capacitors measuring more than a few volts. The residual voltage can affect in-circuit measurements even if it's well below hazardous levels.

Some people use a screwdriver to electrically short the two terminals of a capacitor to discharge it, but this is a hazardous practice. It can produce a big spark and vaporize a chunk out of your screwdriver tip or capacitor terminal. In some circumstances, the capacitor can even explode. It's better to discharge a capacitor through a power resistor (Fig. 2-1). For more details, see the section "Capacitors" in Chapter 7.

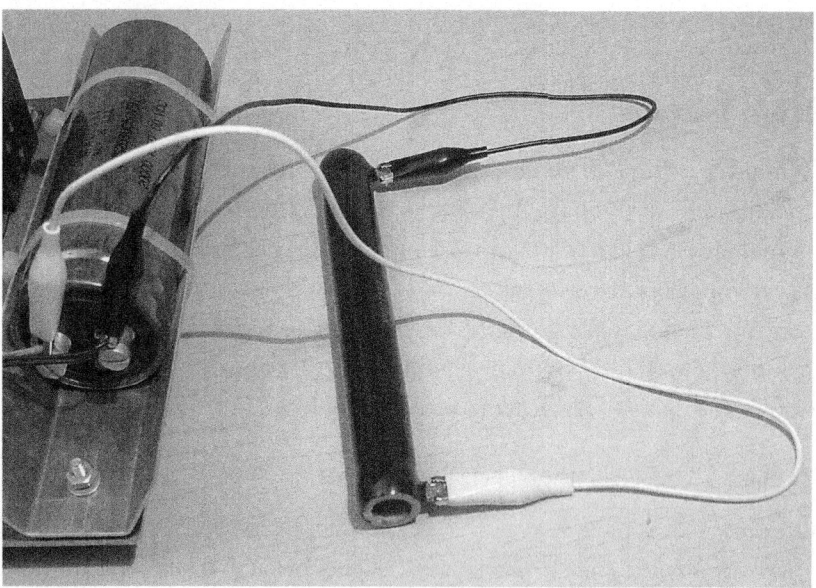

FIGURE 2-1 Discharging an electrolytic capacitor through a resistor.

Inverter Circuits

Inverter circuits convert lower voltages to higher voltages. For example, a voltage inverter is used to provide the several thousand volts needed by a cold-cathode fluorescent lamp (CCFL) backlight in a laptop computer or tablet, even though the device is running on low-voltage batteries. Inverters are used in uninterruptible power supplies for computers, 12-VDC to 120-VAC power inverters for cars, camera flashes, and more.

Don't be fooled into thinking that you're safe just because a device runs on batteries or DC from an AC adapter. Make sure that all power and batteries are removed when working near inverter circuits.

Lasers

Consumer products such as CD players, CD-ROM drives, DVD players, DVD-ROM drives, optical data storage devices, and LaserDisc players use lasers, which can produce vision hazards. Avoid staring directly into them. Devices such as laser pointers and laser levels are more powerful and create a higher hazard.

Solder and Soldering

Soldering irons and molten solder are hot and can burn you or your clothing. Don't lay a hot soldering iron directly on a table; use a soldering iron stand. Molten solder can splash, especially when desoldering, so wear appropriate eye protection.

Solder may contain lead or other hazardous substances. Keep it away from food and drinks. Always wash your hands when you're done working. In addition, soldering may produce hazardous fumes; always solder in a well-ventilated area.

Solvents and Chemicals

Most solvents useful for electronics repair are combustible or flammable, and they are harmful if swallowed. You should avoid breathing fumes and avoid skin and eye contact. In addition, solvents and chemicals should always be stored in a safe place.

Eye Protection

Solder splashes, deflected chemical sprays, exploding components, and flying clipped component leads all pose eye hazards. Always wear appropriate eye protection.

Childproof Your Work Area

An electronics repair workshop is filled with potential hazards, especially for young children. In addition to high voltage, hot soldering irons, and harmful chemicals, your work area likely has sharp tools and electronic components. Colorful components can look like candy and pose a choking hazard if swallowed.

Try to find a work area that's inaccessible to young kids. If that's not possible, make sure that all potential hazards are out of reach when you're not in your work area.

Don't Work Alone

If an accident happens, having another person nearby can make a critical difference. This is especially important when high voltage is involved.

Electronic Waste Disposal

Improper disposal of electronic waste may not directly affect your personal safety, but it can affect the safety of the earth's air and water supplies. Lead and other chemicals leach into the groundwater in landfills or enter the air when burned. Don't throw broken electronic items or e-waste generated from your repair activities in the regular trash. Dispose of them properly through e-recycling programs in your area.

CHAPTER 3

Electronics Basics

If you're new to electronics, this chapter will introduce you to the basics of electronic components, block and schematic diagrams, and electricity.

Introduction to Components

Despite their wide variety of appearances, most electronic components are resistors, capacitors, diodes, transistors, or integrated circuits. Knowing what they look like and what they do is essential for repair work and is the purpose of this section. Specific part identification, part substitution, and typical failure modes are the subject of Chapter 7.

Resistors

Resistors (Fig. 3-1) implement electrical resistance as a circuit element. Resistors are used to reduce current flow, adjust signal levels, bias transistors, and many other things.

Potentiometers

Potentiometers (Fig. 3-2) are variable resistors with three terminals. The two outer terminals have a fixed resistance value between them. The

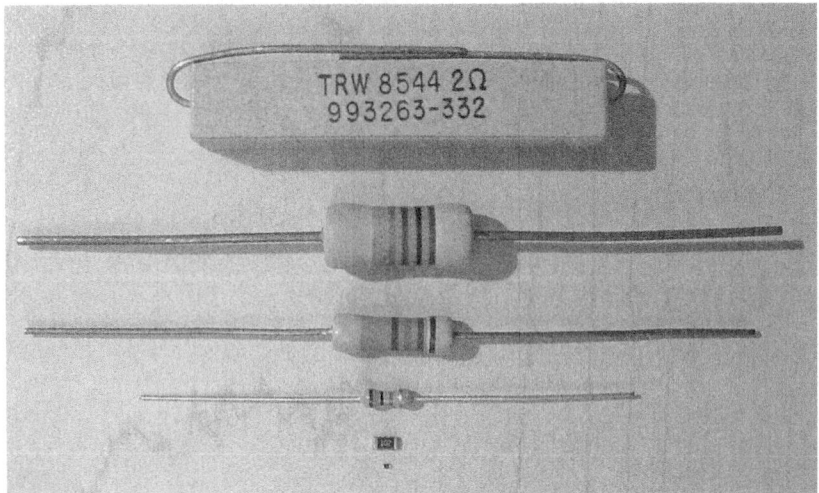

FIGURE 3-1 Variety of resistors.

middle terminal, called the *wiper*, can be adjusted anywhere along the resistance between the two outer terminals.

Potentiometers are used for controls of all types, such as volume and level controls. They can be rotary or sliding. They can have a shaft for a knob or a screwdriver slot for adjustment.

FIGURE 3-2 Variety of potentiometers.

Capacitors

Capacitors (Fig. 3-3), also called *caps*, are charge-storage devices. In power supply circuits, they behave like miniature well-pump pressure tanks: they smooth the flow of current. Capacitors allow high-frequency signals to pass through but block low-frequency signals including DC. This makes them useful in filters, for example, a loudspeaker crossover that only allows high-frequency audio signals to get to the tweeter.

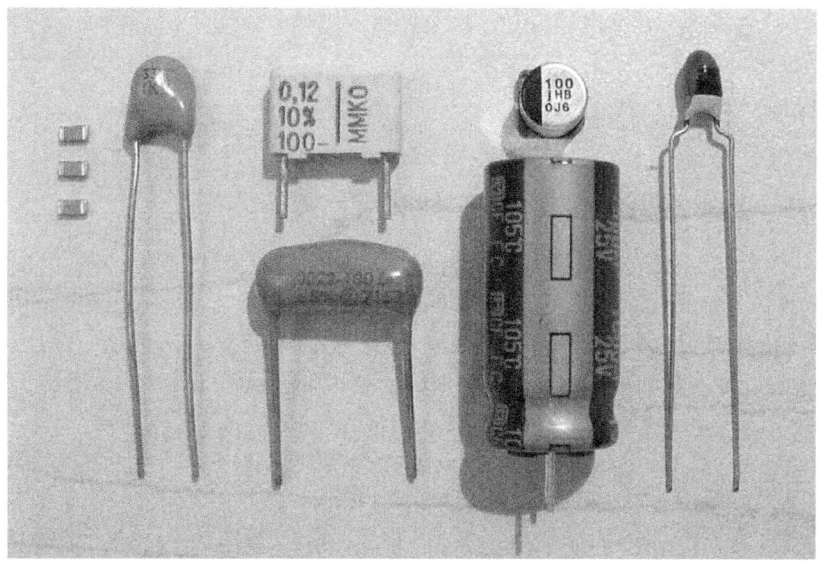

FIGURE 3-3 Variety of capacitors.

Inductors

Inductors (Fig. 3-4), also called *coils*, behave oppositely to capacitors in that they allow low-frequency signals including DC to pass through but block high-frequency signals. They are used less frequently than capacitors because of their relative expense and because they can radiate and pick up magnetic interference.

FIGURE 3-4 Variety of inductors.

Diodes

Most diodes (Fig. 3-5) are used as one-way valves for electric current. In power supply circuits, this one-way-valve behavior is what allows conversion of AC to DC voltage. There are also zener diodes that are used as voltage references and varactor diodes that are used as voltage-dependent capacitors in some high-frequency circuits.

Transistors

Transistors (Fig. 3-6) are used to amplify or switch electronic signals and electrical power. There are many types: NPN and PNP transistors and N-channel and P-channel MOSFETs (metal-oxide–semiconductor field-effect transistors), to name a few. Most have three terminals.

Electronics Basics | 13

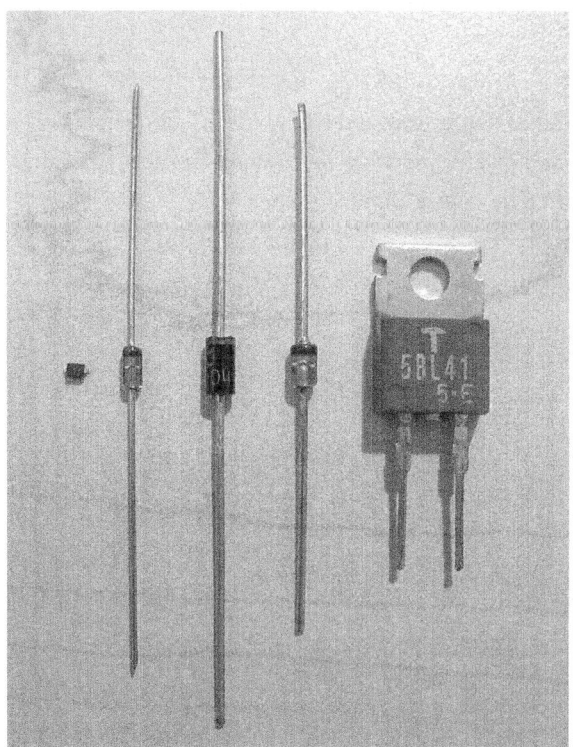

FIGURE 3-5 Variety of diodes.

FIGURE 3-6 Variety of transistors.

Integrated Circuits

Integrated circuits (Fig. 3-7), also called *ICs,* are used in virtually all electronic equipment and have revolutionized the world of electronics. ICs can contain as many as a billion transistors in a single package. ICs range in size from a grain of rice to several square inches.

A special case of ICs are three-pin voltage regulators. They look just like transistors; the only way to tell the two apart is by the numbers printed on the package.

FIGURE 3-7 Variety of integrated circuits.

Transformers

Transformers (Fig. 3-8) convert a higher AC voltage to a lower AC voltage (step-down) or a lower AC voltage to a higher AC voltage (step-up). They also provide electrical isolation between the input and output. Both of these traits come in handy in power supply circuits, where you'll almost always find them.

Electronics Basics | 15

FIGURE 3-8 Variety of transformers.

Relays

Relays (Fig. 3-9) are electrically operated switches. Two terminals are for an electromagnetic coil inside the relay package; the others are switch terminals. When the coil is energized with electricity, it activates the mechanical switch inside. Many relays have multiple switches inside, controlled by the same electromagnetic coil. Relays allow a relatively low control voltage to control a high-powered switch circuit. For example, a 5-VDC control signal from a microprocessor could be used to close the switch inside a relay connecting 120 VAC to a motor.

FIGURE 3-9 Variety of relays.

Introduction to Schematics

Many types of diagrams are used to show the details of electronic circuits. The most common are *block diagrams* and *schematic diagrams*. Schematic diagrams are also called *circuit diagrams* or just *schematics*.

Block Diagrams versus Schematics

Block diagrams (Fig. 3-10) are intended to show how a device works by showing the important functional systems and how they connect to the other systems. Block diagrams are very helpful when you need to understand how a device works in order to troubleshoot it.

Schematic diagrams (Fig. 3-11) show the details of how all the electronic components are electrically connected. Well-drawn schematics are drawn in such a way as to aid in understanding how the circuit works and may include normal DC voltages and even oscilloscope waveforms for important circuit nodes.

Electronics Basics | 17

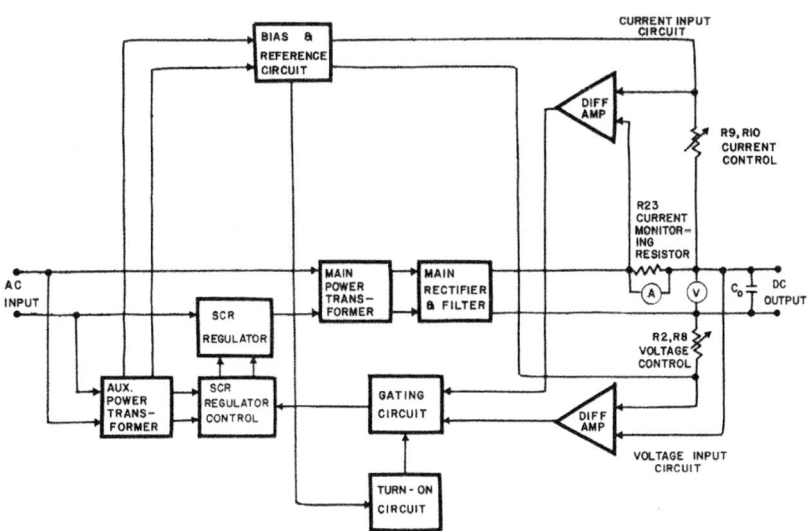

FIGURE 3-10 Example of a block diagram.

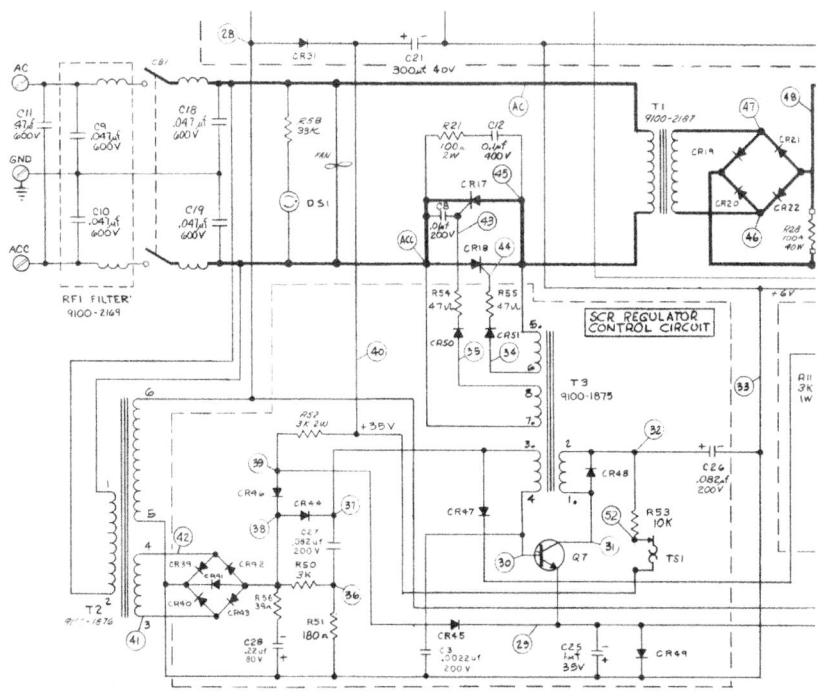

FIGURE 3-11 Example of a schematic diagram.

To make sense of a schematic, you need to understand the symbols used to represent each of the types of common components.

Component Symbols

Symbols vary by country and manufacturer but will always be consistent within a schematic. The symbols in Fig. 3-12 are those normally used in the United States.

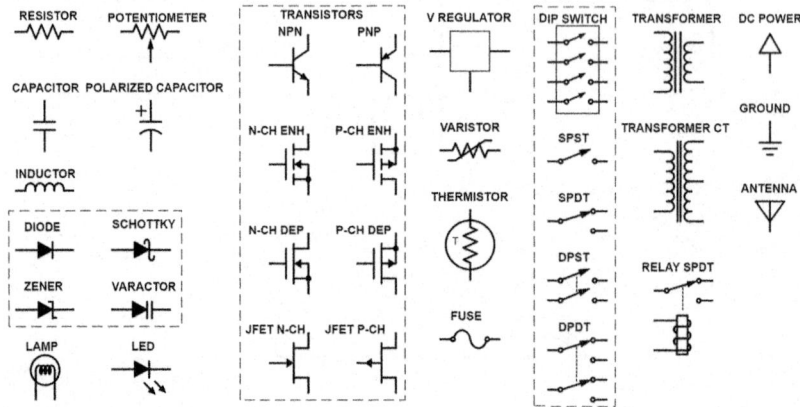

FIGURE 3-12   Component symbols.

There are two conventions for showing connections between wires (Fig. 3-13). The first convention uses a dot to show a connection. If two wires cross and there is no dot at the intersection, there is no connection.

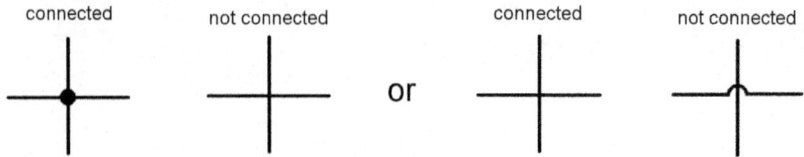

FIGURE 3-13   Wire connections in schematics.

The second convention uses half-circles to show no connection. If two wires cross and there is no half-circle, there is a connection.

Connections to ground and power can be shown using individual ground and power symbols for each circuit node, or they can be shown all wired together with a single ground or power symbol attached. The first method makes the diagram easier to comprehend. If the second method is used, you can take a black marker to highlight all the ground wiring and a red marker to highlight all the power wiring to improve clarity.

Introduction to Electricity

Having a basic understanding of electricity helps when it comes to repairing electronics, especially when troubleshooting is required to figure out what's wrong.

Basic Units of Electricity

The four basic units of electricity are voltage, current, resistance, and power. One way to help understand these electrical quantities is by thinking about water and plumbing (Table 3-1).

TABLE 3-1 Electrical Quantities versus Water Quantities

Electrical Quantity	Equivalent Water Quantity
Voltage (volts, V)	Water pressure (pounds per square inch)
Current (amperes or amps, A)	Water flow (gallons per minute)
Resistance (ohms, Ω)	Pipe resistance
Power (watts, W)	Pressure × flow (horsepower, hp)

Voltage and current are easiest to understand: voltage is like water pressure and current is like water flow.

Electrical resistance is analogous to a water pipe's resistance to flow. Imagine trying to fill a swimming pool from your hose spigot with a regular garden hose. Then think about how much longer it would take if the hose were the diameter of spaghetti. This is because of the higher resistance of the spaghetti-sized hose. In both cases, the water pressure (voltage) is the same, but in the case of the higher resistance, the flow (current) is reduced.

Ohm's Law

In electricity, this same relationship holds. It's called *Ohm's law*:

$$\text{Current} = \text{voltage}/\text{resistance}$$

This says that you can double the current (water flow) by doubling the voltage (water pressure) or by halving the resistance (pipe resistance). Ohm's law can be rearranged to calculate any of the three quantities if you know the other two.

Power Formulas

In electricity, power is calculated using this equation:

$$\text{Power} = \text{voltage} \times \text{current}$$

In terms of water, this means that you need both water pressure and flow to hose the mud off your car. Having lots of pressure but little flow is like trying to wash it with a squirt gun. Having lots of flow but little pressure is like gently pouring buckets of water over the car. You need pressure *and* flow to effectively wash your car.

When resistance is involved, this equation can be combined with Ohm's law and rearranged to produce two alternate equations for power:

$$\text{Power} = \text{voltage}^2/\text{resistance}$$

$$\text{Power} = \text{current}^2 \times \text{resistance}$$

These equations come in handy when you're trying to figure out the wattage rating you need for a resistor if you know the maximum voltage or current to which it will be subjected.

Batteries in Series

This is probably obvious to anyone who has replaced batteries in a flashlight, but the total voltage of batteries in series is the total of the voltages of the individual cells. Rechargeable battery packs for cordless power tools are composed of series-connected cells, usually 1.2 V each. So a 12-V battery pack would have 10 cells in series and an 18-V battery pack would have 15 cells. Series-connected batteries (Fig. 3-14) connect the positive terminal of one battery to the negative terminal of the next battery.

Occasionally, batteries are connected in parallel. This means that the positive terminals are connected together and the negative terminals are connected together. This configuration doesn't increase the voltage, but it does increase the amount of run time.

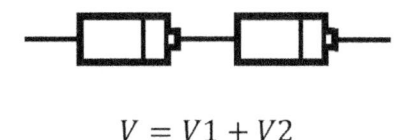

$$V = V1 + V2$$

FIGURE 3-14 Batteries in series.

Resistors in Series and Parallel

It's common to find resistors wired in series or parallel (Fig. 3-15). In series, the values add, just as with batteries. In parallel, the combined resistance is the product over the sum. In other words, you multiply all the individual resistor values together and divide that number by the sum of all the individual resistor values.

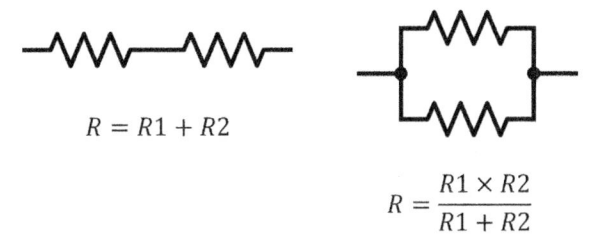

$$R = R1 + R2$$

$$R = \frac{R1 \times R2}{R1 + R2}$$

FIGURE 3-15 Resistors in series and parallel.

This comes in handy when you're looking to build a resistor that you don't have out of resistors that you do have for troubleshooting or repair.

It also comes in handy for building a cheap dummy load for amplifier or power supply testing. This will be covered in Chapter 4.

Capacitors in Parallel

Capacitors behave oppositely from batteries and resistors: you have to connect them in parallel for the values to add (Fig. 3-16). This is a common arrangement in power supply circuits and makes it easier to keep a low profile on the PC board.

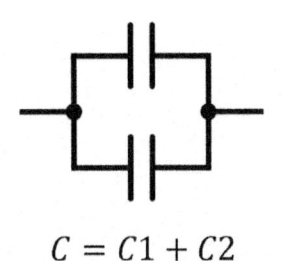

$$C = C1 + C2$$

FIGURE 3-16 Capacitors in parallel.

It is possible to use capacitors in series, but it's not commonly done. Series-wired caps use the same equation as parallel-wired resistors: the combined value is the product over the sum.

AC versus DC

Most people are familiar with the terms *DC voltage* and *AC voltage* but may not understand the difference. DC stands for *direct current*, and AC stands for *alternating current*. DC voltage remains constant over time. Batteries are good examples of sources of DC voltage.

AC voltage alternates polarity, usually on a periodic basis. For household power in the United States, the polarity changes from plus-to-minus-to-plus 60 times a second. That's called 60 cycles per second or 60 hertz (Hz). The voltage follows the shape of a sine wave over time.

Light bulbs and heating elements work equally well with AC or DC power. AC motors rely on an alternating voltage to operate and won't work with DC. In contrast, most electronics need DC to work. Fortunately, it's practical to convert AC to DC with just a handful of parts.

WHY DO WE USE AC?

You might wonder why AC is used for providing power to our homes. The short answer is that AC makes it relatively easy to convert from one voltage to another by using transformers. Transformers don't work for DC.

Around the turn of the last century, when decisions were being made about power distribution over long distances, there were two competing factions. Thomas Edison was promoting a DC distribution system and Nikola Tesla an AC distribution system. The technically superior AC distribution prevailed.

By converting the AC to very high voltage for transmission (as high as a million volts), the same amount of power can be transmitted over thinner wires with lower power losses. That means cheaper electricity!

CHAPTER 4

Equipment and Supplies

Repairing electronics requires some specialized supplies and equipment. You may already have some of what you'll need. The usefulness of many of the items in this chapter depends on the type of repair work you'll be doing. Start small and build your equipment empire as you go.

The real value of this chapter is that it lets you discover something to make repairs faster, easier, cheaper, or better. This chapter is organized by basic supplies, hand tools, and equipment from basic to advanced.

Basic Supplies

Electrical Contact Cleaner Sprays

Electrical contact cleaner sprays are the closest thing I've found to a repair kit in a can. I've had great success in bringing equipment back to life by spraying switches, audio connectors, and potentiometers (pots) with the appropriate spray. I use CAIG DeoxIT D5 (Fig. 4-1) for switches and connectors and DeoxIT Fader F5 (Fig. 4-2) for pots.

Electrical contacts oxidize over time from exposure to the air, resulting in intermittent or bad connections. In mild cases, bad connections produce scratchy volume and tone controls in audio equipment. In more extreme cases, the equipment just doesn't work. I've had many pieces of test equipment that were completely nonfunctional until treated with contact spray.

The trick with contact sprays is to get the chemical into the contacts without getting it everywhere else. The CAIG FLEX-TIP helps with this. After spraying, work the switch or pot back and forth over its entire range many times to distribute the chemical and scrub the contacts. Clean up the excess and let dry.

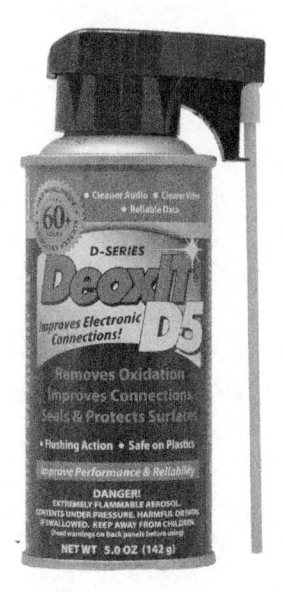

FIGURE 4-1 CAIG DeoxIT D5.
(Courtesy of CAIG Laboratories, Inc.)

FIGURE 4-2 CAIG DeoxIT Fader F5.
(Courtesy of CAIG Laboratories, Inc.)

Air Duster

Air duster (Fig. 4-3) is like having compressed air in a can. It's useful for blowing dust and debris out of hard-to-reach crevices such as computer keyboards.

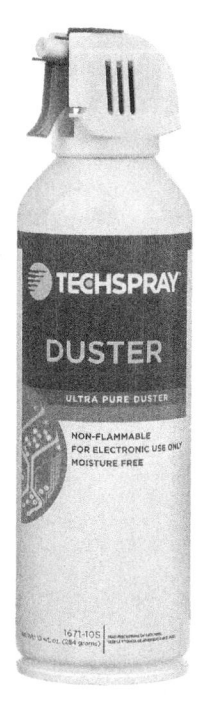

FIGURE 4-3 TechSpray Air Duster.
(Courtesy of Techspray www.techspray.com.)

Freeze Mist

Freeze Mist (Fig. 4-4) is an important tool in tracking down temperature-related intermittent problems. Spraying it on an electrical component will instantly chill it to subzero temperatures. If you have a problem that only seems to appear after warm-up, you can selectively freeze suspected misbehaving parts to see if that makes the problem go away.

Alternately, if you have a problem that disappears after warm-up, you can use Freeze Mist to try to find the part that misbehaves when cold. Freeze Mist can cause ice to form, which then melts to form water, so a misbehaving part found in this way may be a false alarm if it is moisture sensitive.

Freeze Mist is also useful in finding intermittent problems that don't seem to be temperature related. For example, the freezing action can cause enough thermal contraction to make a bad solder joint act up.

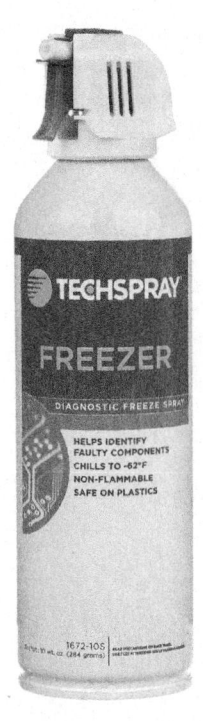

FIGURE 4-4 TechSpray Freezer.
(Courtesy of Techspray www.techspray.com.)

Solder

See Chapter 5 for information on choosing the right solder.

Solder Wick

Solder wick is used for removing solder to free electronic components for replacement. See Chapter 5 for more information.

Heat-Shrink Tubing

Heat-shrink tubing (Fig. 4-5) is used to insulate and protect electrical connections or components. When heated with a heat gun or lighter, it shrinks to about half its original diameter (Fig. 4-6). The length stays about the same.

FIGURE 4-5 Variety of heat-shrink tubing and tape.

FIGURE 4-6 Before and after heat shrinking.

Heat-shrink tubing comes in rolls or sections that you cut to the length you want. It's available in a wide range of diameters and colors, including clear. You'll want to have a variety of sizes on hand.

Some heat-shrink tubing is lined with a layer of hot-melt adhesive that seals the inside of the tubing for better environmental resistance. Unlike electrical tape, which you can apply after the fact, you'll need to plan ahead with heat-shrink tubing and slide it on the wire before soldering the connection. If you make a mistake, you'll have to start from scratch and desolder and resolder the connection.

You can also buy adhesive-lined heat-shrink tape that you use like electrical tape then heat to shrink and bond.

Electrical Tape

Electrical tape is used to insulate and protect electrical connections or components. It is inferior to heat-shrink tubing for most electronics applications. The adhesive used in electrical tape is notorious for turning to goo after a few years, so buy the best you can find.

Electrical tape is available in colors (Fig. 4-7). One good use for it is color-coding to keep track of which screws go where when taking TVs apart. See the section "Taking It Apart" in Chapter 8.

FIGURE 4-7 Electrical tape in colors.

Heat-Sink Compound

Heat-sink compound is used to help conduct heat from transistors or integrated circuits (ICs) to their heat sinks. See the section "Heat-Sinked Devices" in Chapter 5 for more information on heat-sink compound and the less messy alternative, thermal pads.

Nail Polish

Nail polish is useful for painting the tops of potentiometers inside a device after you've adjusted the pots (Fig. 4-8). This will reduce the chances that the pots will vibrate to a different setting in use.

FIGURE 4-8 Nail polish on trim pot.

Wire Ties

Wire ties (Fig. 4-9), also known as *tie wraps*, *cable ties*, and *zip ties*, are a great way to hold things together. In electronics, they're used for bundling wires, securing cables to assemblies, and many other applications.

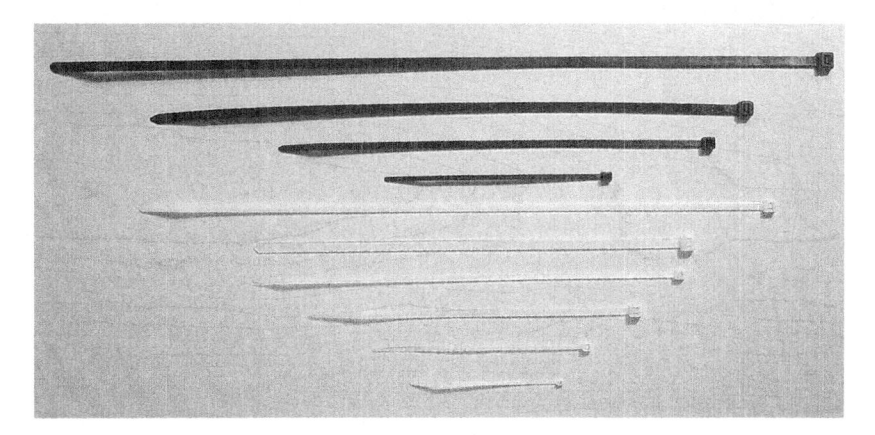

FIGURE 4-9 Variety of wire ties.

Most wire ties are made of nylon and are available in a variety of sizes and colors. They tighten with a ratcheting action. Some have a release tab, but most need to be cut off for disassembly and then replaced with a new one.

Solvents

Solvents are regularly used in electronics repair to remove flux after soldering, clean old heat-sink compound from semiconductors, and remove adhesive residue, oil, and grease. Because solvents can damage plastics and other materials, the best approach is to start with something mild. Isopropyl alcohol (Fig. 4-10) is the starting solvent of choice for electronics. Buy 99% or higher isopropyl. Don't use the 70% isopropyl commonly found in drugstores; it has too much water in it. Always test your solvent on a cotton swab in a small, inconspicuous area before diving in.

FIGURE 4-10 TechSpray isopropyl alcohol.
(Courtesy of Techspray www.techspray.com.)

Equipment and Supplies | 33

> **SAFETY NOTE**
>
> Most solvents useful for electronics repair are combustible or flammable and harmful if swallowed. You should avoid breathing fumes and avoid skin and eye contact. Solvents should always be stored in a safe place.

Glues

In repair work, you'll eventually need glue. Forget about superglue; epoxy is king (Fig. 4-11). Superglue forms a strong bond but has very low shear strength. This means that the parts bonded will resist direct pulling but not off-angle stress. Epoxy, by contrast, has tremendous structural strength. Epoxy is also better at filling voids between parts.

FIGURE 4-11 Loctite Quickset Epoxy. *(Courtesy of Henkel Adhesive Technologies.)*

You'll need to mix resin and hardener, so it's a little more work, but it's worth it. Epoxy products are available with setting times as short as 1 minute. You'll sacrifice some strength with the short-setting-time formulations, so look at the strength number on the package as well as the setting time.

Epoxy repair putty (Fig. 4-12) combines the adhesiveness of epoxy with the ability to add structure. It's a great way to fill in gaps or holes or to reinforce a repair to reduce the likelihood that it will break again.

34 | The Electronics Repair Cookbook

For hard-to-bond plastics such as polypropylene and polyethylene, try Loctite Super Glue Plastics Bonding System with Activator (Fig. 4-13).

FIGURE 4-12 Loctite Epoxy Repair Putty.
(Courtesy of Henkel Adhesive Technologies.)

FIGURE 4-13 Loctite Plastics Bonding System.
(Courtesy of Henkel Adhesive Technologies.)

Clip Leads

Clips leads (Fig. 4-14) come in handy for experimentally connecting components to a circuit while debugging. They're useful for discharging power supply capacitors through a resistor (see "Electrolytic Capacitors" in Chapter 2). With clip leads, you can temporarily extend the length of the ground wire of your scope probe so that you don't have to hunt for ground within a few inches of each measurement point. You can use clip leads to temporarily convert voltmeter needle-tipped probes to clips. Clip leads are a convenient temporary antenna when you're working on stereo receivers. Get a mix of clip sizes, lengths, and colors so that you're ready for every clip-lead need.

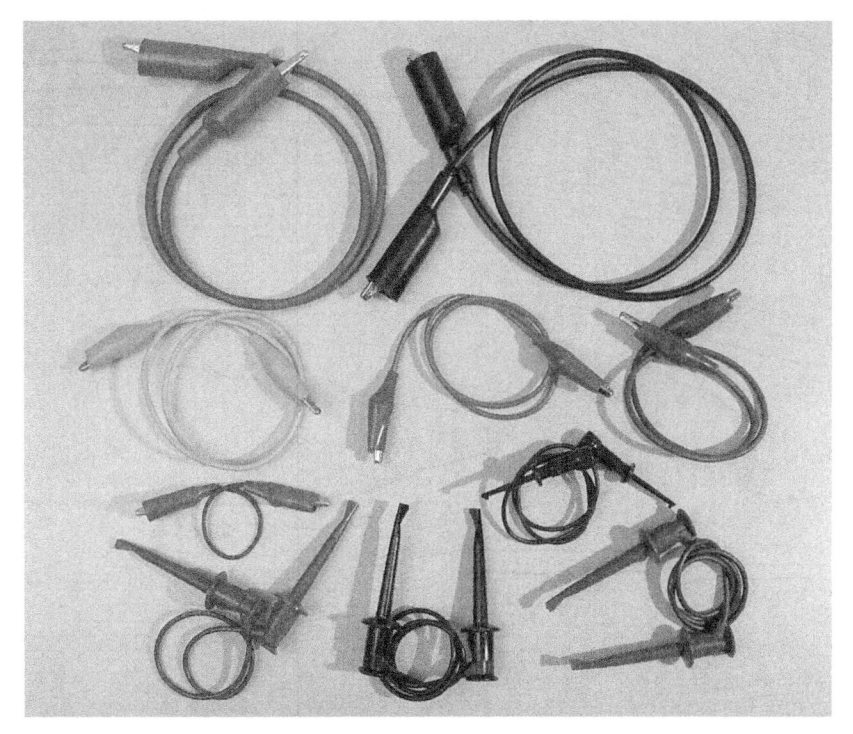

FIGURE 4-14 Variety of clip leads.

USE MAGNETS TO MAKE CLIP LEADS EVEN HANDIER

You can use magnets to hold the alligator clips of clip leads to flashlight battery terminals and other steel objects (Fig. 4-15). This solves the problem of not having a good place to clip onto. Put magnets between batteries to stack them when you need more voltage than a single cell provides. For this to work, you'll need strong, electrically conductive magnets. Neodymium magnets are a good choice. If the magnet doesn't look like metal, it's probably nonconductive and won't work. Keep a handful of suitable magnets with your clip leads.

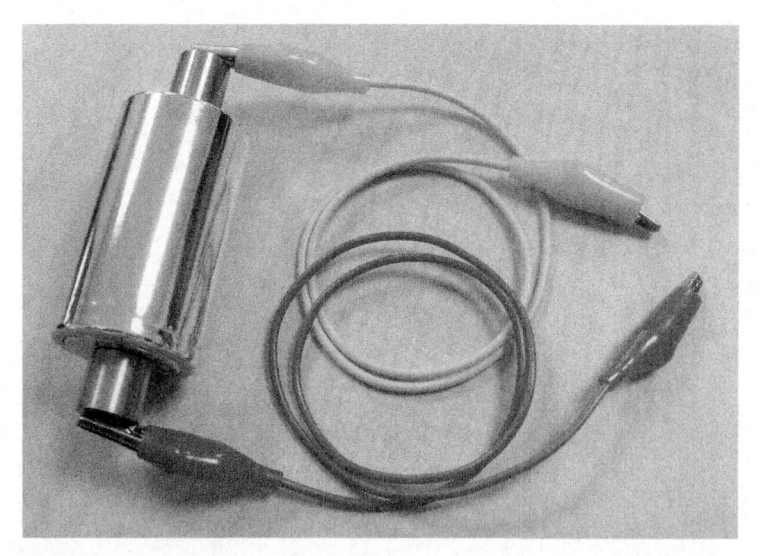

FIGURE 4-15 Magnetic clip-lead trick.

Crimp Connectors

Crimp connectors (Fig. 4-16), also called *solderless connectors,* are commonly used in car stereo installation, where you may not want a hot soldering iron near leather upholstery. Crimp connectors also have their uses in general electronics repair, particularly if you want a ring or spade lug on the end of a wire.

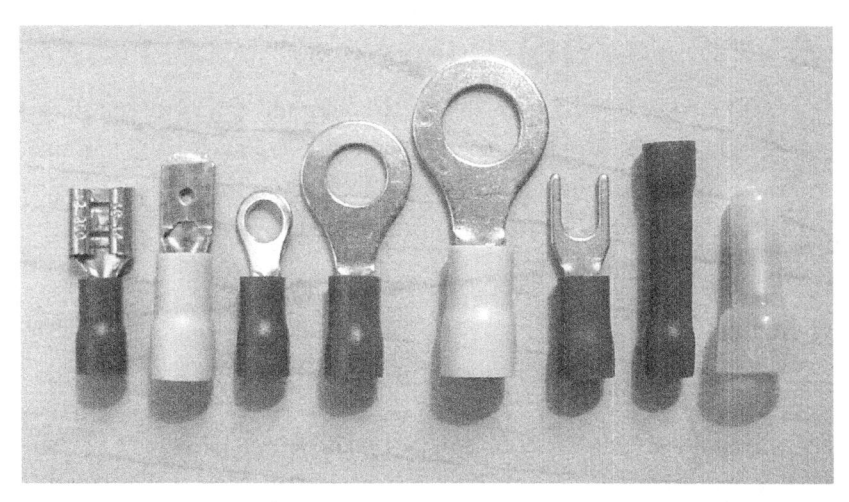

FIGURE 4-16 Variety of crimp connectors.

My favorite crimp connector for splicing wires together is the closed-end connector. With a closed-end connector, you twist the stripped wire ends together and then crimp the connector on. This gives good wire-to-wire contact as well as wire-to-connector contact.

The key to success with crimp terminals is to have the right size connector for the gauge of wire. If you try to use a connector that's too large for the wire gauge, it will not make a good connection and is likely to slide off. Terminals are generally color-coded according to the industry standard, as shown in Table 4-1. If you don't know the wire gauge, use a crimper/stripper to figure it out by stripping some insulation, or just choose the smallest crimp connector that fits the wire. Be sure to use the *pull test* after each crimp.

TABLE 4-1 Crimp Connector Color Coding

Wire Gauge	Standard Crimp-Connector Color
22–18	Red
16–14	Blue
12–10	Yellow

> **TIP**
>
> Don't tin the ends of your wires with solder unless you intend to solder them to the crimp connector. Solder is a relatively soft alloy and will readily cold flow under pressure, causing compression connections such as unsoldered crimp connectors to eventually loosen.

Hand Tools

Needle-Nose Pliers

It's hard to think of an electronics repair project that doesn't use needle-nose pliers (Fig. 4-17).

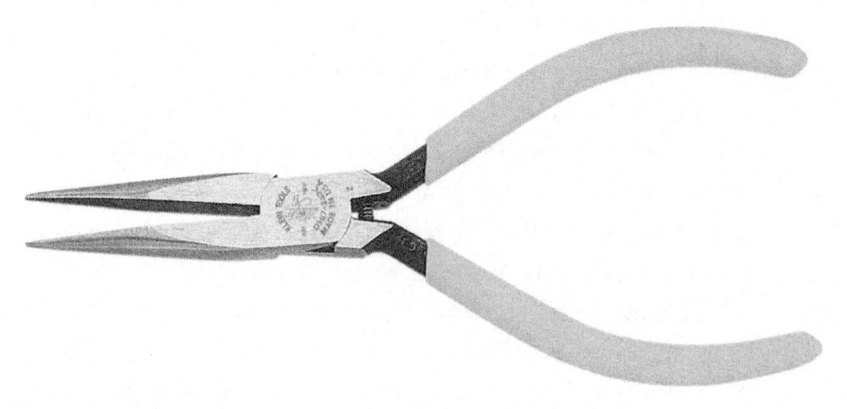

FIGURE 4-17 Klein needle-nose pliers D307-51/2C. *(Courtesy of Klein Tools.)*

Wire Cutters

Wire cutters come in a variety of styles. It's useful to have a pair that can handle thicker wires and cables (Fig. 4-18). It's also useful to have a pair of lightweight flush cutters (Fig. 4-19) for precision work.

FIGURE 4-18 Klein diagonal cutters D202-6C.
(Courtesy of Klein Tools.)

FIGURE 4-19 Klein flush cutters D275-5.
(Courtesy of Klein Tools.)

Wire Stripper

Although you can use wire cutters to strip off insulation from the end of a wire, you're likely to cut into the wire itself and nick it or cut off a few strands of stranded wire. It is better to use a wire stripper (Fig. 4-20).

FIGURE 4-20 Klein wire stripper 11046.
(Courtesy of Klein Tools.)

X-Acto Knife

A precision razor knife finds many uses in electronics repair. I like the X-Acto Gripster (Fig. 4-21) because its rear blade release makes it easier and safer to replace blades. It has an antiroll design and comes with a safety cap. The no. 11 blade is standard. Uses include cutting printed-circuit (PC) board traces, scraping the solder mask off a trace to expose bare copper, removing corrosion, and enlarging holes.

Precision Screwdriver Set

For the tiny screws found in many of today's electronic devices, you'll need screwdrivers to match (Fig. 4-22). Laptops and cameras are prime examples of products loaded with small screws.

Large screws are usually forgiving of using a screwdriver of the incorrect size. With miniature screws, you need to more carefully match the screwdriver tip to the screw head to avoid stripping the screw head.

Equipment and Supplies | 41

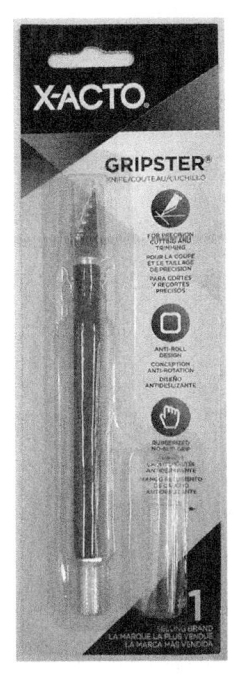

FIGURE 4-21 X-Acto Gripster knife with a no. 11 blade.

FIGURE 4-22 Wiha precision slotted/Phillips screwdriver set 26197. *(Courtesy of Wiha Tools.)*

JIS (Japanese Industrial Standard) Screwdriver Set

JIS screws look a lot like Phillips screws, but they're different (see the following sidebar). If you use a Phillips screwdriver to loosen or tighten them, you may end up with stripped screw heads.

JIS screws are often marked by a dimple in the head, but not always. They're common in personal electronics from laptops to digital cameras. JIS screwdrivers (Fig. 4-23) can be used with both JIS and Phillips screws with no damage.

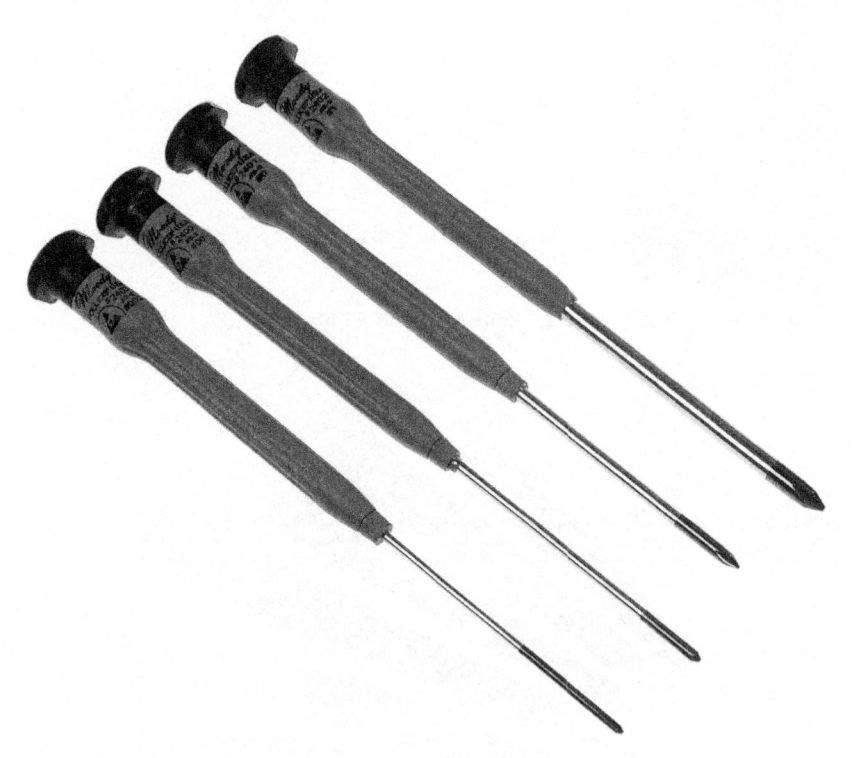

FIGURE 4-23   JIS screwdriver set. *(Courtesy of iFixit.)*

WHAT'S THE DIFFERENCE BETWEEN JIS AND PHILIPS SCREWS?

In 1936, Henry F. Phillips patented the Phillips screw. It was a great solution for automobile production lines because it was designed to "cam-out" once a certain torque was reached, to prevent overtightening of the screw. Also, unlike slotted-head screws, the Phillips self-centering design allowed operators to engage the tip of the driver into the screw head quickly and easily.

In Japan, engineers developed their own cross-point design (Fig. 4-24). Like a Phillips screwdriver, the Japanese cross-point drivers have self-centering and quick tool-and-screw engagement. The key difference is that the JIS design allows torque and overtightening to be controlled by the operator, not the head of the screw.

FIGURE 4-24 JIS screw with dimple.

Because of the different profiles (Fig. 4-25), a conventional Phillips screwdriver tip won't fully seat in a JIS screw head. One reason is that the corner radius at the cross section on a JIS screw head is smaller than that of a Phillips tipped screwdriver. Also, most JIS screws have a shallower cavity and since a Phillips screwdriver has a longer tip design, it won't fit all the way into a JIS screw. As a result, a Phillips tip may not grip the sides of the JIS screw properly and will most likely cause the operator to damage the screw.

By contrast, a JIS screwdriver fits both JIS and Phillips screws perfectly. If you don't know what type of screw you've got, use a JIS driver because it's universal for both screw types.

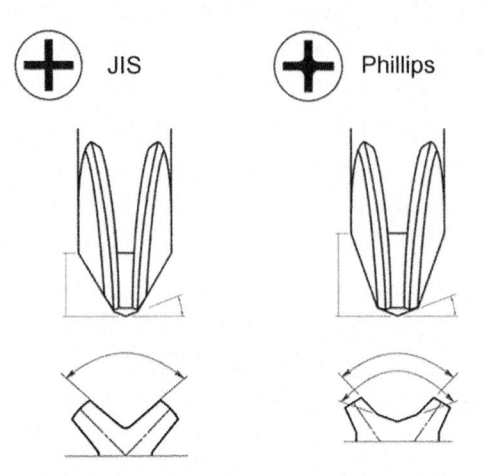

FIGURE 4-25 JIS versus Phillips screwdriver tips.

Security-Bit Set

Many electronic products such as video games and television set-top boxes use security screws to make them difficult to open. Security-bit sets (Fig. 4-26) contain a selection of bits to fit the most common security screw types, for use with a ratcheting screwdriver.

FIGURE 4-26 Klein 32-piece tamperproof bit set 32525. *(Courtesy of Klein Tools.)*

Solder Sucker

A solder sucker is used for removing solder to free electronic components for removal and replacement. See Chapter 5 for more information.

Tweezers

Precision tweezers (Fig. 4-27) are indispensable for replacing surface-mount components on PC boards. They're also useful for handling miniature screws during disassembly and reassembly.

FIGURE 4-27   Wiha ESD safe precision tweezers 44501. *(Courtesy of Wiha Tools.)*

Hemostats

Originally designed for surgeons, hemostats combine the agility of small needle-nose pliers with clamping action (Fig. 4-28). The ratcheting clamp is released by moving the rings together with a slight sideways force. This makes them useful as a third hand for clamping or part holding while soldering.

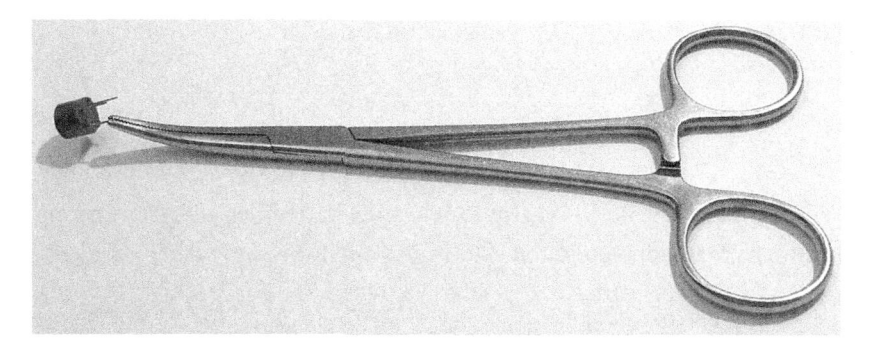

FIGURE 4-28   Hemostats holding a part.

Pin Vise

A pin vise is a handle for holding a drill bit. It has an adjustable collet that accommodates a range of drill bit sizes. The one I use is the Starrett 166A (Fig. 4-29), which holds a drill bit from zero to 0.040 inch in diameter. I normally use a 0.033-inch bit.

The main use is for cleaning out a component lead hole where some solder still remains after desoldering and further attempts at desoldering are likely to damage the copper. A secondary use is drilling a virgin hole in copper to add a new component, relocate a component, or use a component of a different size. If you're just adding a hole or two, a few twists of the pin vise is easier than getting out a drill.

FIGURE 4-29 Starrett 166A pin vise with 0.033-inch bit. *(With permission of the Starrett Company.)*

Dental Pick

A dental pick (Fig. 4-30) is great for cleaning battery leakage from spring terminals in battery-operated electronics and for removing flux and corrosion from PC boards.

Crimper

If you plan on using solderless crimp connectors, you'll need a crimper for them (Fig. 4-31).

FIGURE 4-30 Dental pick.

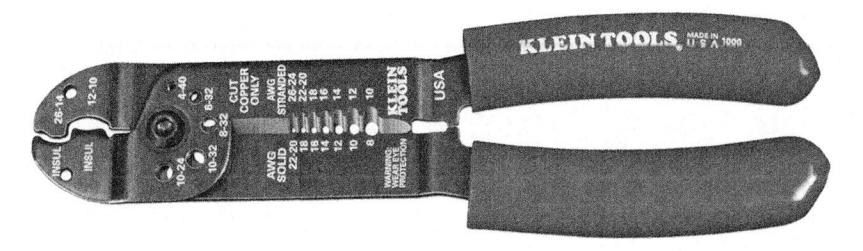

FIGURE 4-31 Klein multipurpose tool 1000. *(Courtesy of Klein Tools.)*

Strain-Relief Pliers

You can use ordinary pliers to compress power cord strain-relief grommets, but strain-relief pliers (Fig. 4-32) work a whole lot better. They give you a better grip and don't damage the grommet or surrounding area. It's surprisingly satisfying to replace a power cord without chewing up the grommet.

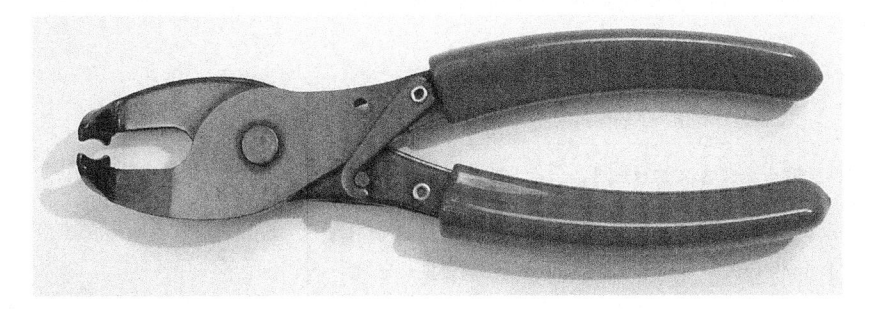

FIGURE 4-32 Strain-relief pliers.

Nut-Driver Set

A hollow-shaft nut-driver set (Fig. 4-33) may seem like a luxury if you already own a socket or wrench set, but nut drivers can reach hex screws at the bottom of narrow tubular channels that would otherwise be impossible to access. The hollow shafts let you remove nuts from bolts that extend a few inches past the nut, a problem with a regular socket set.

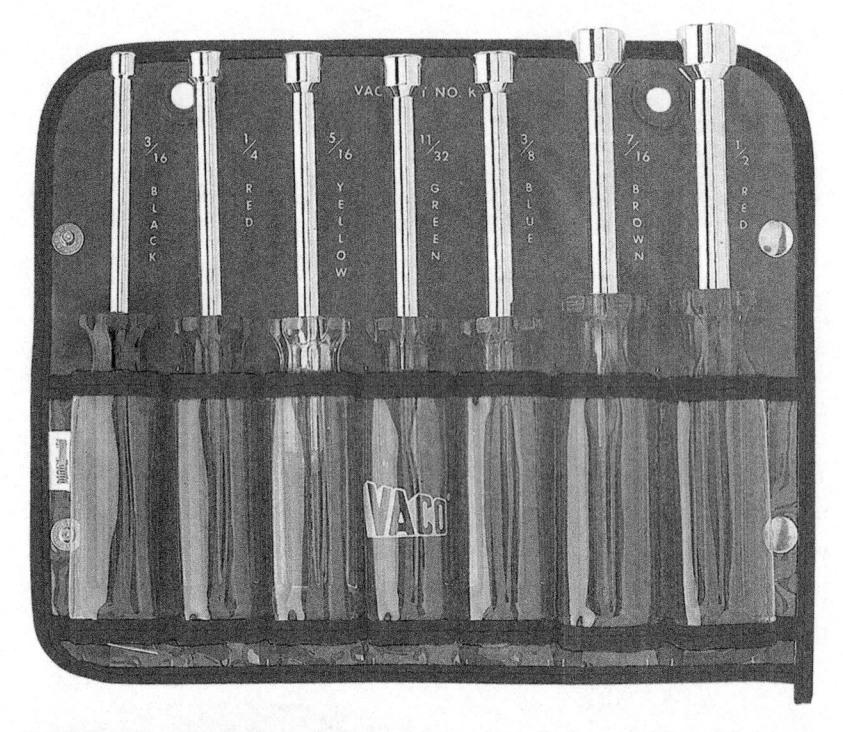

FIGURE 4-33 Klein hollow-shaft nut-driver set K7. *(Courtesy of Klein Tools.)*

Basic Equipment

The basic equipment category includes relatively inexpensive items that you're likely to use on most electronics repair projects.

Soldering Iron/Soldering Station

A bare-bones soldering iron may suffice if you only expect to solder a couple of joints a few times a year. If you expect to solder more often, you'll want a soldering station. A soldering station generally incorporates a soldering iron holder and a sponge holder. The wet sponge is used for cleaning the hot iron tip. Adjustable temperature is a common and useful feature. The soldering irons in soldering stations usually have easily replaceable tips in different sizes and types. This is a key feature that lets you cover a range of soldering tasks from miniature surface-mount chips to large metal shields. The Weller WLC100 (Fig. 4-34) is an economical 40-watt soldering station with sponge holder and adjustable temperature. It accommodates the Weller ST Series of tips (Fig. 4-35).

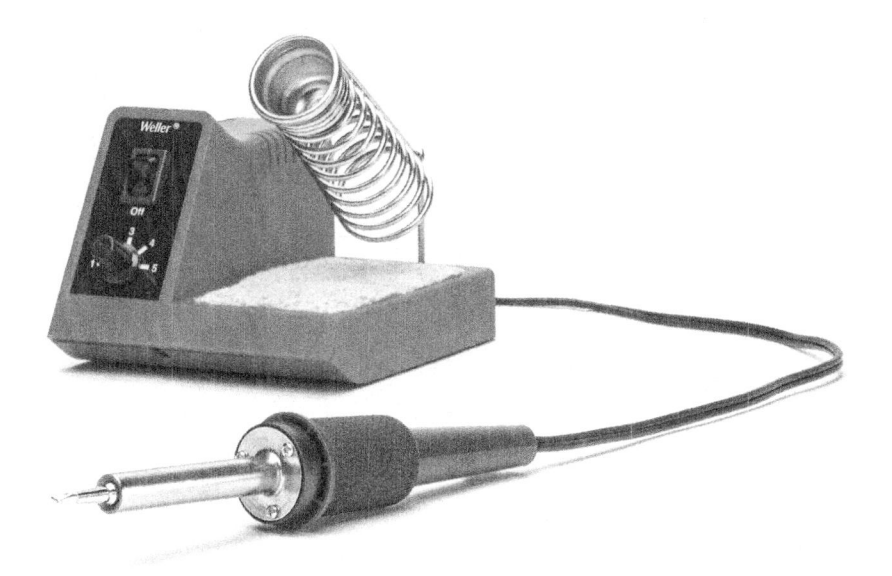

FIGURE 4-34 Weller WLC100 soldering station. *(Courtesy of Weller.)*

If you're willing to invest a little more money, look at the Weller WE1010. It's a 70-watt iron with true temperature control. This means that you can control the actual temperature at the tip, not just the wattage of the iron. It accommodates the Weller ET Series of tips. The WE1010 also has a high-temperature silicone cord to the iron that resists melting from accidental contact with the soldering iron tip. See Chapter 5 for more information on using soldering equipment.

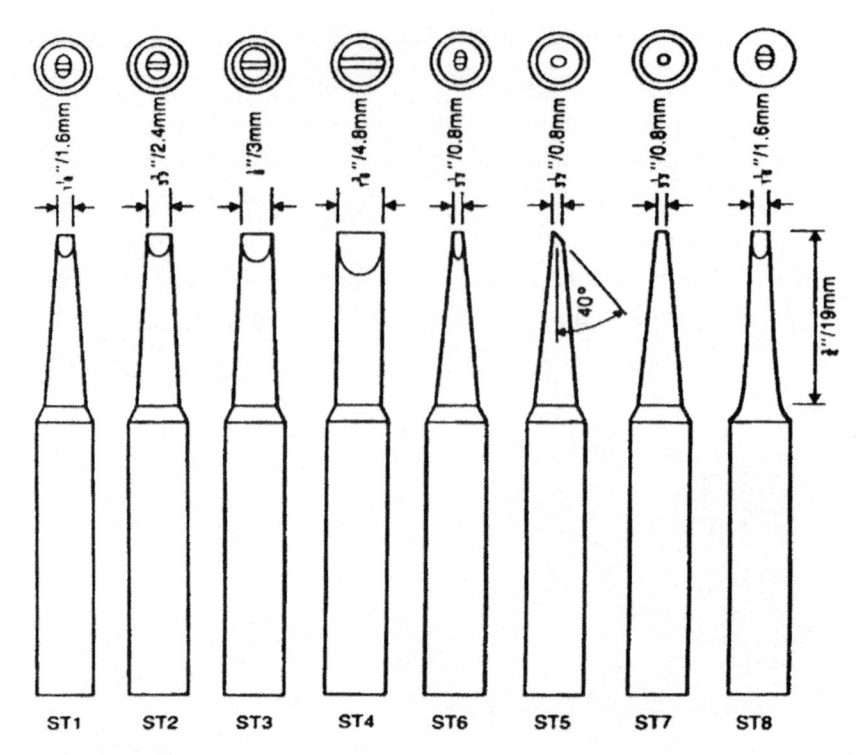

FIGURE 4-35 Weller ST Series tips for the WLC100 soldering station. *(Courtesy of Weller.)*

Digital Multimeter

A digital multimeter (Fig. 4-36) is the single most useful diagnostic tool for electronics repair. It lets you measure direct-current (DC) and alternating-current (AC) voltages, resistance, current, and more. Two important features that are not always found on inexpensive multimeters are diode check and audible continuity. Make sure you get a meter with them. Other useful features found on better multimeters are frequency and capacitance measurement. See Chapter 6 for information on how to use a digital multimeter.

Equipment and Supplies | 51

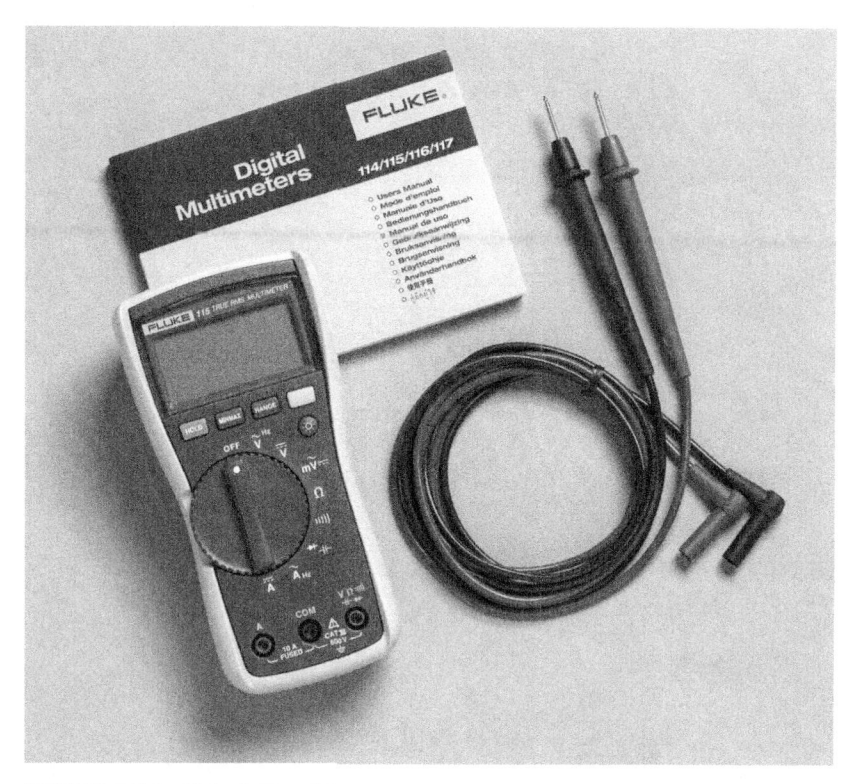

FIGURE 4-36 Fluke 115 multimeter.
(Reproduced with permission, Fluke Corporation.)

Magnifiers: OptiVISOR/Magnifier Lamp

Today's electronics are so small that magnification is frequently needed to perform repairs. There are many inexpensive magnification options from which you can choose. I like the OptiVISOR binocular headband (Fig. 4-37) because it provides stereoscopic magnification and is comfortable to wear. You can choose from six different levels of magnification. For example, the DA-7 model provides 2.75 times magnification and is a good choice for microelectronics. Another option is a magnifier lamp (Fig. 4-38). Choose a model in which the lens is large enough to allow stereoscopic viewing.

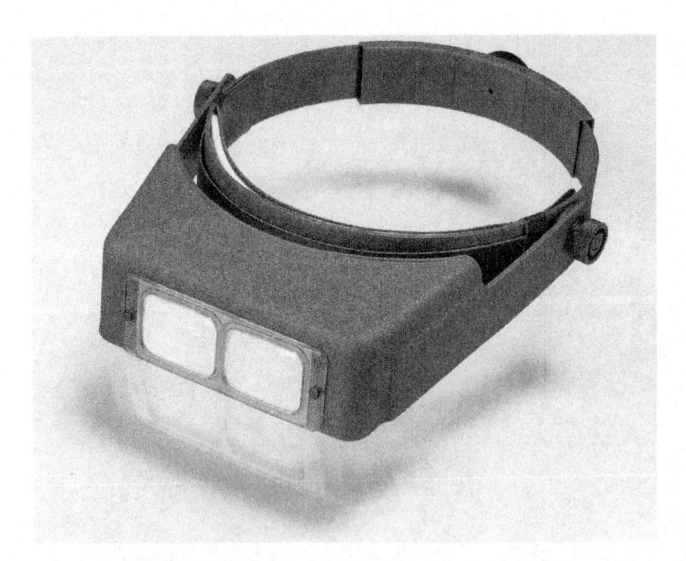

FIGURE 4-37 OptiVISOR headband magnifier. *(Courtesy of Donegan Optical.)*

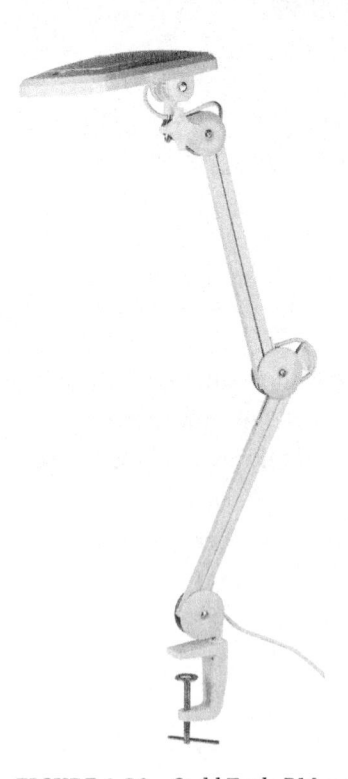

FIGURE 4-38 Stahl Tools RM magnifier lamp. *(Courtesy of Parts Express.)*

Antistatic Mat and Wrist Strap

Many of today's semiconductors are easily damaged by static electricity, also called *electrostatic discharge* (ESD). Frequently, the damage weakens the part, causing it to fail prematurely. The best defense is to use an antistatic mat (Fig. 4-39) and wrist strap (Fig. 4-40) when handling static-sensitive circuit boards and parts.

To use an antistatic mat, place the mat on your work surface. Place the equipment you're repairing on the mat. Connect the first cord from the snap on the mat to the equipment metal chassis using the alligator clip. Put the antistatic wrist strap on your wrist (with the metal part touching your skin), and snap the second cord onto the wrist strap. Attach the alligator clip of the second cord to the antistatic mat. You're now ready to work.

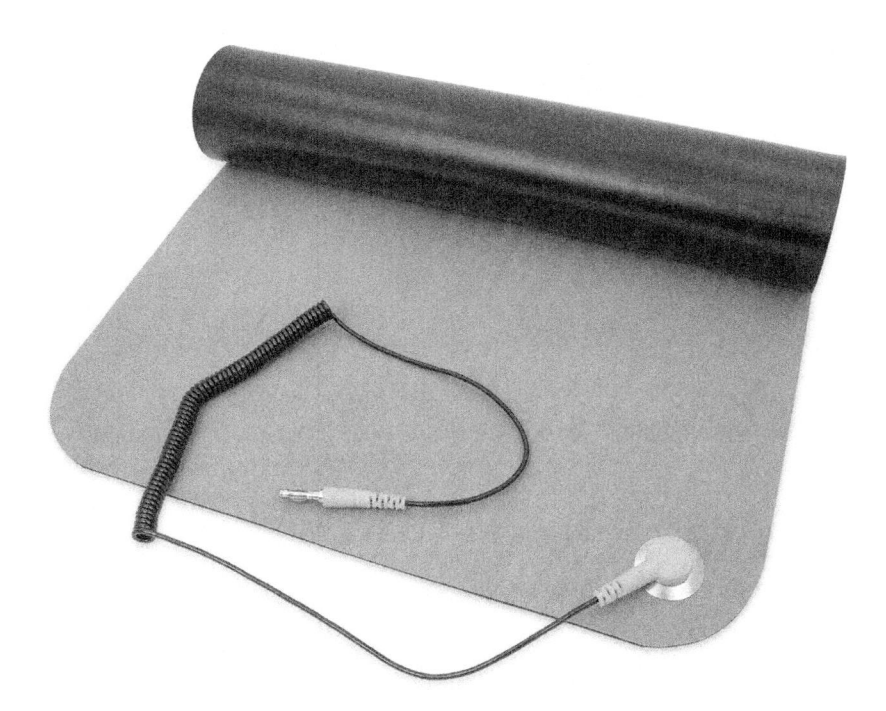

FIGURE 4-39 Antistatic mat. *(Courtesy of iFixit.)*

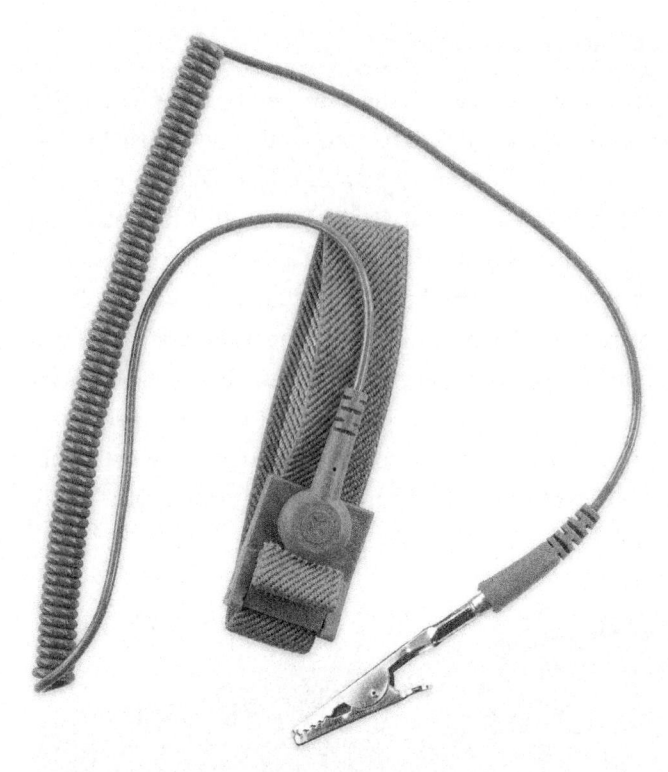

FIGURE 4-40 Antistatic wrist strap. *(Courtesy of iFixit.)*

> **TIP**
>
> So-called wireless and cordless antistatic wrist straps are sold online. These do not work, so don't waste your money on them.

Intermediate Equipment

The intermediate equipment category includes relatively inexpensive items that you're likely to want to use on some electronics repair projects.

Heat Gun

A heat gun is useful for shrinking heat-shrink tubing and for isolating temperature-related intermittent problems. It's also handy for softening adhesives and reflowing hot glue. With care, you can use it to warm adhesive labels for removal and reuse. The Weller 6966C (Fig. 4-41) is a lightweight heat gun that provides both hot and cold air and has a small nozzle to let you precisely direct the airflow.

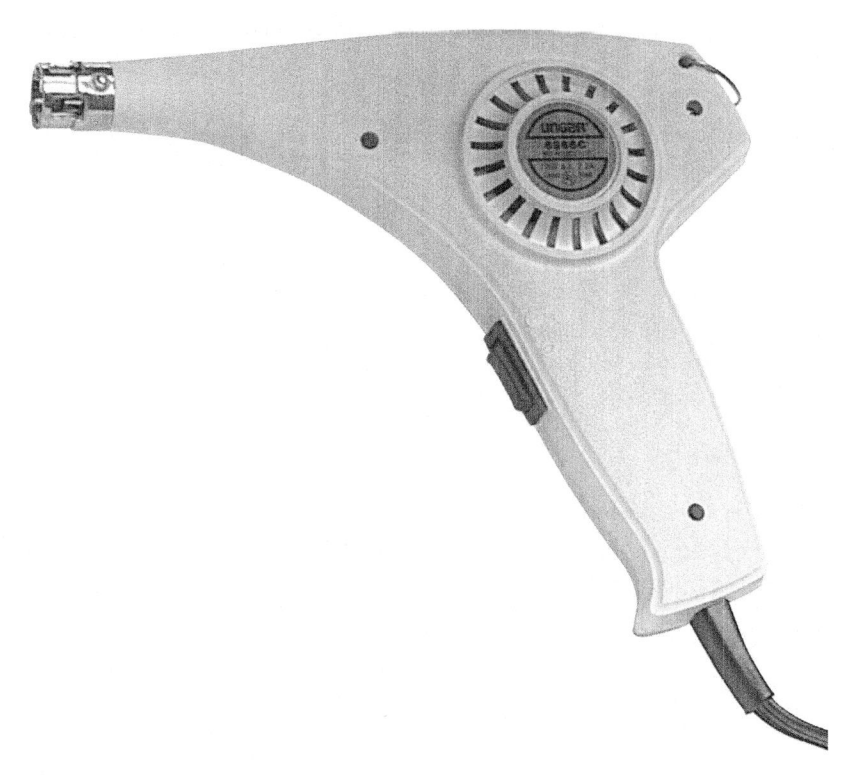

FIGURE 4-41 Weller 6966C heat gun. *(Courtesy of Weller.)*

Digital Calipers

A ruler or tape measure is fine for large measurements, but for microelectronics, you need a more precise way to measure small dimensions. Digital calipers (Fig. 4-42) are the tool of choice. Examples of what you might want to measure with calipers include

- Size and lead spacing of electronic components
- Diameter of component leads

- Diameter and pitch (thread spacing) of screws
- Diameter of wire to determine its gauge

The Mitutoyo 500-196-30 AOS Absolute Digimatic Calipers provides English and metric measurements for inside, outside, and depth.

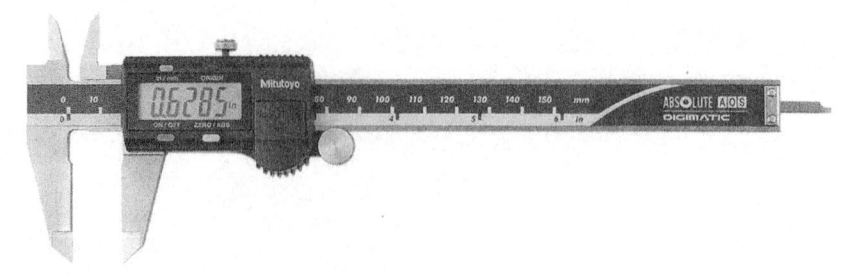

FIGURE 4-42   Mitutoyo 500-196-30 digital calipers. *(Courtesy of Mitutoyo America.)*

Panavise

Have you ever wished you had a third hand? Panavise makes an array of work-holding tools that make soldering and measurement easier. The Panavise Model 350 Multipurpose Work Center (Fig. 4-43) can hold objects up to 9 inches wide. The jaws can be reversed to provide zero closure. A single knob controls the head movement through three planes: 90° tilt, 360° turn, and 360° rotation.

Dummy Loads

A dummy load is a power resistor used to test power supplies, audio amplifiers, and other devices. A power supply may appear to work just fine with no load connected, but it malfunctions once you start to draw current. Similarly, an audio amplifier might seem to work with no load or an 8-ohm load connected, but it malfunctions with a 4-ohm load. Or it might work at low volume levels but distort at high volume levels. You could use sets of speakers to do your testing, but you'd run the risk of damaging them. You'd also run the risk of disturbing your neighbors with high-volume level testing. The solution is dummy loads.

FIGURE 4-43 Panavise Model 350 multipurpose work center. *(Courtesy of PanaVise Products, Inc.)*

You can buy a power resistor with the resistance and power rating you want (Fig. 4-44) or build your own dummy load by soldering together resistors that you can get cheaply at a hamfest or from an electronics surplus dealer. You'll need to use the equations in the "Resistors in Series and Parallel" section of Chapter 3.

Suppose that you want a 100-watt, 4-ohm load. You could build that out of ten 40-ohm, 10-watt resistors wired in parallel. Because 40 ohms isn't a standard resistance value, the closest value is 39 ohms, so you'll end up with 3.9 ohms.

For power supply testing, it's advantageous to use series-wired resistors in a homemade dummy load. With series-wired resistors, you can connect to the two ends of the resistor string or anywhere in between. This lets you choose a resistance value that is any multiple of the resistors you use in the string. My homemade dummy load (Fig. 4-45) uses eight 10-ohm, 22-watt resistors in series. This lets me test power supplies by choosing any multiple of 10 ohms that I want, up to 80 ohms.

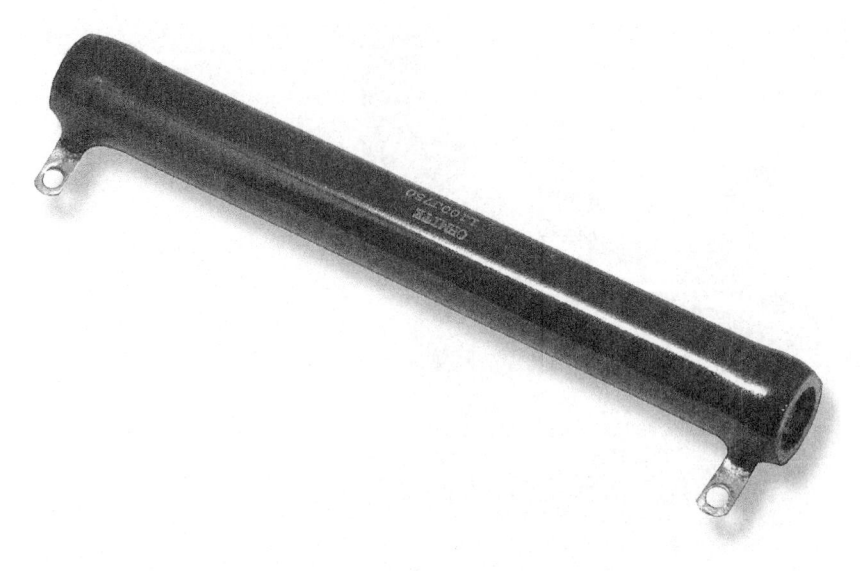

FIGURE 4-44 Ohmite 270 Series power resistor.

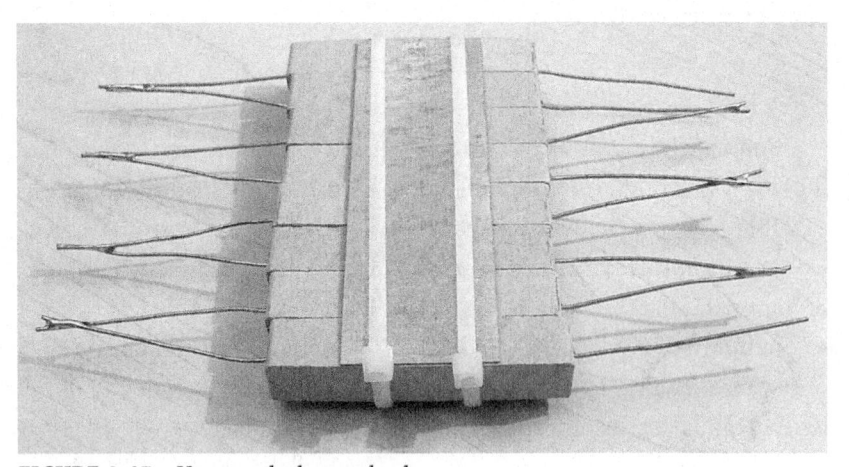

FIGURE 4-45 Homemade dummy load.

Advanced Equipment

The advanced equipment category includes equipment that's more expensive than that in the basic and intermediate categories, but the equipment in this category is also more powerful and can make the difference between being able to make a particular repair or not.

Power Supply

Adjustable DC power supplies are useful for powering equipment such as car stereos that require a source of DC power to run. You can run battery-operated equipment with a power supply when you don't have batteries on hand or the rechargeable battery pack is low. If the equipment being repaired comes with its own external power supply brick and you suspect that the brick is bad, you can use an adjustable power supply to power the equipment and confirm that the only problem is the brick. Similarly, if the internal power supply circuit of a device is down, you can use an adjustable DC power supply to see if the only problem is the internal supply.

When choosing a power supply, the main considerations are voltage range, maximum current, and number of outputs. Usually, the maximum current specification applies to all output voltage settings, but in some power supplies, the maximum current depends on the output voltage. Some power supplies have a digital display to show voltage and current, some have analog meters, and some have no display at all.

Adjustable current limiting is a useful feature that lets you limit the maximum current going to an external device. This helps protect your equipment from damage. Set the current limit for what's needed for normal operation.

Another factor to consider is whether a power supply is linear or switching. Linear supplies use a hefty power transformer and linear regulator. They produce a cleaner output but are larger, heavier, and less efficient (more heat). Switching supplies are more likely to have some high-frequency noise present on the output, and they may radiate high-frequency energy electromagnetically. They are smaller, lighter, more efficient, and less expensive than a comparable linear supply.

The B&K Precision 1671A Triple-Output DC Power Supply (Fig. 4-46) provides 0–30 volts at 5 amperes. It also has fixed 12- and 5-volt outputs. It provides adjustable current limiting and has digital voltage and current displays. It's a switch-mode supply.

FIGURE 4-46 B&K Precision 1671A adjustable DC power supply. *(Courtesy of B&K Precision.)*

ESR Meter

One of the most common component failures in today's electronics is an electrolytic capacitor. Bad caps often show visible signs of failure, such as bulging tops or leaking electrolyte, but in many cases, they look perfectly normal.

The best way to check for bad caps is an ESR meter (Fig. 4-47). An ESR meter measures the equivalent series resistance of a cap. This is a much better indicator of capacitor health than simply measuring capacitance. The measured ESR value is compared with a value on a chart to determine if it's good or not (Fig. 4-48).

Another advantage of an ESR meter over an ordinary capacitance meter is that it allows in-circuit testing so that you don't need to remove the cap from the circuit board. This is because ESR meters use a low-voltage signal that will not turn on semiconductors.

Capacitors can hold a high-voltage charge for minutes or hours that can damage test equipment if not properly discharged before measurement. The Peak Atlas ESR70 automatically carries out a controlled discharge procedure before measuring the capacitance and ESR.

Equipment and Supplies | 61

FIGURE 4-47 Peak Atlas ESR70. *(Courtesy of Peak Electronic Design Ltd.)*

	10V	16V	25V	35V	63V	160V	250V	400V	630V
4.7µF	42.0Ω	35.0Ω	29.0Ω	24.0Ω	20.0Ω	16.0Ω	13.0Ω	11.0Ω	8.5Ω
10µF	20.0Ω	16.0Ω	14.0Ω	11.0Ω	9.3Ω	7.7Ω	6.3Ω	5.3Ω	4.0Ω
22µF	9.0Ω	7.5Ω	6.2Ω	5.1Ω	4.2Ω	3.5Ω	2.9Ω	2.4Ω	1.8Ω
47µF	4.2Ω	3.5Ω	2.9Ω	2.4Ω	2.0Ω	1.60Ω	1.30Ω	1.10Ω	0.85Ω
100µF	2.0Ω	1.60Ω	1.40Ω	1.10Ω	0.93Ω	0.77Ω	0.63Ω	0.53Ω	0.40Ω
220µF	0.90Ω	0.75Ω	0.62Ω	0.51Ω	0.42Ω	0.35Ω	0.29Ω	0.24Ω	0.18Ω
470µF	0.42Ω	0.35Ω	0.29Ω	0.24Ω	0.20Ω	0.16Ω	0.13Ω	0.11Ω	0.09Ω
1000µF	0.20Ω	0.16Ω	0.14Ω	0.11Ω	0.09Ω	0.08Ω	0.06Ω	0.05Ω	0.04Ω
2200µF	0.09Ω	0.08Ω	0.06Ω	0.05Ω	0.04Ω	0.04Ω	0.03Ω	0.02Ω	0.02Ω
4700µF	0.04Ω	0.04Ω	0.03Ω	0.02Ω	0.02Ω	0.02Ω	0.01Ω	0.01Ω	0.01Ω
10000µF	0.02Ω	0.02Ω	0.01Ω	0.01Ω	0.01Ω	0.01Ω	0.01Ω	0.01Ω	0.00Ω
22000µF	0.01Ω	0.01Ω	0.01Ω	0.01Ω	0.00Ω	0.00Ω	0.00Ω	0.00Ω	0.00Ω

Remember, lower ESR is better.

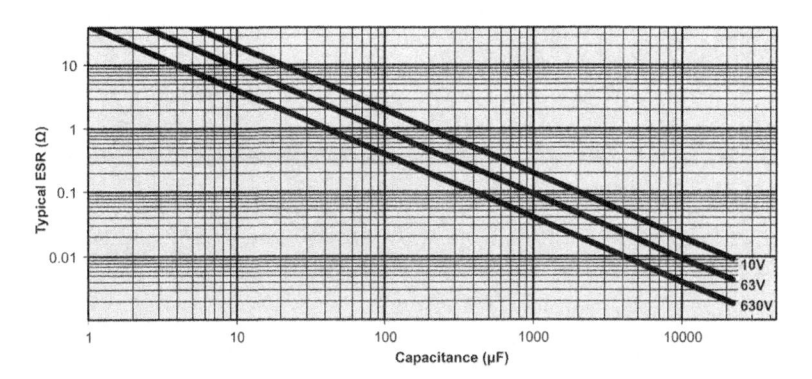

FIGURE 4-48 Peak Atlas ESR70 chart. *(Courtesy of Peak Electronic Design Ltd.)*

Oscilloscope

An oscilloscope (Fig. 4-49), also called a *scope*, lets you "see" an electrical signal on a display. More precisely, it shows the voltage versus time. This provides much more information than a voltmeter, but oscilloscopes are also a lot more expensive than voltmeters and require more skill to operate. For advanced troubleshooting, a scope is a necessity.

When choosing a scope, the main considerations are bandwidth and number of channels. Bandwidth is the maximum frequency that the scope can accurately display. Number of channels is the number of signals that the scope can display simultaneously.

Another factor to consider is whether the scope is analog or digital. Virtually all new scopes are digital, but there are plenty of used analog scopes on the market. Digital scopes can store data, let you capture and display one-time events, and offer useful features such as amplitude and frequency measurement, but they can also produce misleading displays because of *aliasing* and usually require more skill to operate. See Chapter 6 for how to use an oscilloscope.

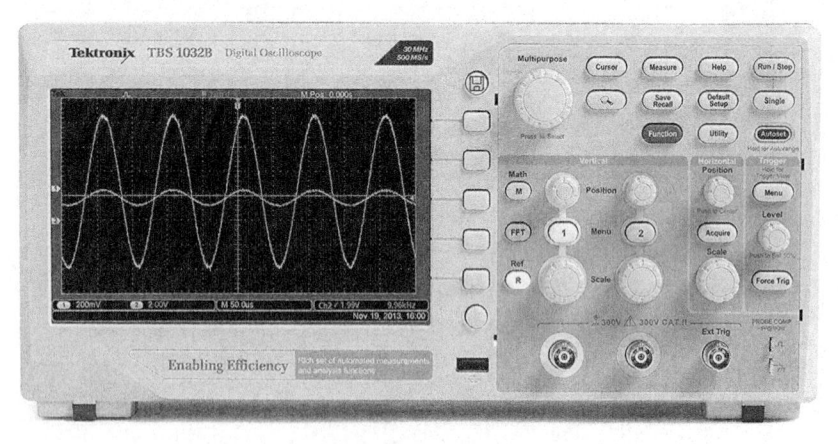

FIGURE 4-49 Tektronix TBS1032B two-channel 30-MHz digital storage oscilloscope. *(Copyright © Tektronix. Reprinted with permission. All Rights Reserved.)*

WHAT ABOUT PC OSCILLOSCOPES?

A PC oscilloscope (or USB oscilloscope) is generally a palm-sized device that connects to a personal computer via a USB cable. PC scopes use the display, computing power, and user-interface capabilities of the connected PC to reduce cost. Some PC scopes cost less than $100.

Despite the affordability of a PC scope, many users prefer a traditional oscilloscope. A few areas where inexpensive PC scopes may fall short include a lack of AC coupling mode, a lack of analog offset capability, and low resolution and bandwidth.

The PicoScope 2206B two-channel 50-MHz PC oscilloscope (Fig. 4-50) provides an AC coupling mode, analog offset capability, and a built-in function generator.

FIGURE 4-50 PicoScope 2206B two-channel 50-MHz PC oscilloscope. *(Image is copyrighted by and used with permission of Pico Technology.)*

Function Generator

A function generator (Fig. 4-51) is a flexible test signal generator. Most can provide sine waves, square waves, triangle waves, and pulse trains over a wide range of frequencies, voltages, and DC offsets. Function generators are especially useful for debugging audio gear in conjunction with an oscilloscope. With a triangle wave supplied to the input of an audio amplifier, debugging becomes a matter of following the signal path through the amp circuitry and looking for waveform anomalies with the scope.

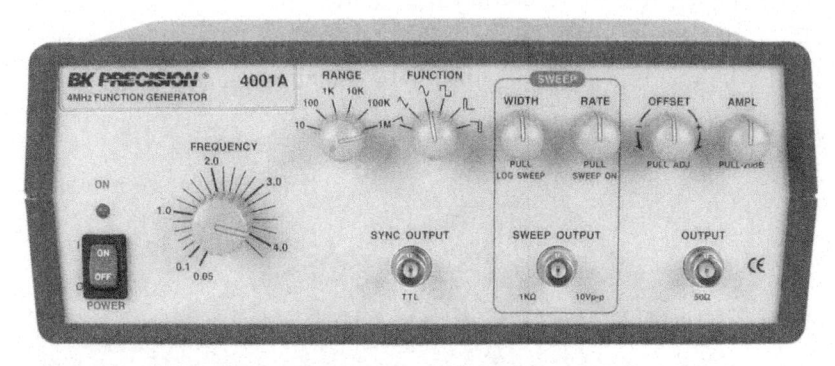

FIGURE 4-51 B&K Precision 4001A function generator. *(Courtesy of B&K Precision.)*

Isolation Transformer

An isolation transformer (Fig. 4-52) electrically isolates the 120 VAC provided to a device from the electrical mains. You plug the isolation transformer into the electrical outlet in your workshop and then plug the device you're repairing into the outlet on the isolation transformer. The purpose of isolation transformers is isolate you from the normally ground-referenced high voltages in a device.

It's important to know that isolation transformers don't fully eliminate the hazard of high voltage. Isolation transformers isolate a device from earth ground. Without an isolation transformer, you could be shocked by accidentally touching a single high-voltage point in a device

FIGURE 4-52 B&K Precision 1604A isolation transformer.
(Courtesy of B&K Precision.)

because you're standing on earth ground, completing the circuit. With an isolation transformer, you could still be shocked by touching two points having a high voltage difference. For this reason, you should always keep one hand behind your back when working on high-voltage (>50 volts) devices. This prevents your opposite hand from completing the electrical circuit and passing current through your heart.

When choosing an isolation transformer, the main consideration is power handling. Another consideration is whether it has a built-in fuse or circuit breaker. Some isolation transformers provide an adjustable output voltage feature. This is useful to verify the operation of equipment for the full range of possible operating conditions, such as during power brownouts.

CHAPTER 5

Soldering and Desoldering

Knowing how to solder and desolder is a necessary skill for electronics repair. With some knowledge and a little practice, everyone can learn to solder well enough for basic repair work. This chapter provides the knowledge. You will have to do the practicing on your own. Take advantage of the many tutorial videos online that show soldering and desoldering in action to hone your skills.

Soldering

Proper Tip Size and Temperature

Chapter 4 recommends an adjustable-temperature soldering station that can accommodate a range of tip sizes and styles (Fig. 5-1).

Choosing the right tip for the job is mostly a matter of size matching. This means using a needle-point tip when soldering tiny surface-mount chip resistors and caps and using a large tip when soldering physically large connections.

Soldering and desoldering operations subject components and the PC board to potentially damaging high temperatures. There are several ways to reduce the chances of causing damage. While it might seem like using a smaller tip and lower temperature setting would reduce the potential for damage, the opposite is usually true. To reduce the chance

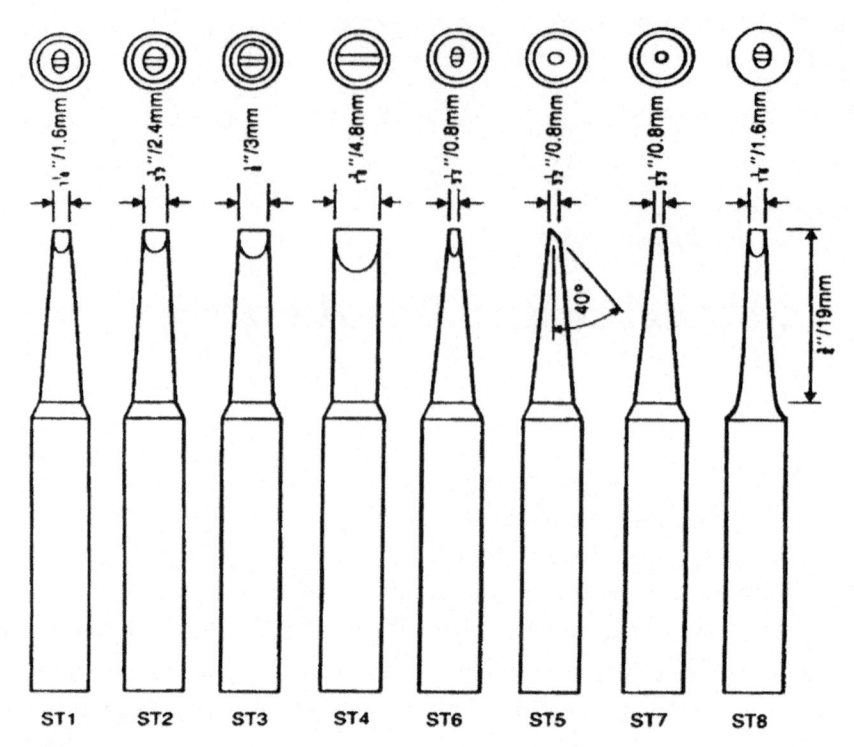

FIGURE 5-1 Weller ST Series tips for a WLC100 soldering station. *(Courtesy of Weller.)*

of damage, you want to reduce the time that you're heating the connection. If the tip is too small or the temperature setting is too low, you'll be spending too much time on the connection, allowing heat to flow where it's more likely to cause damage, such as into the body of a component. If you can't achieve a good solder joint in a few seconds, there's a good chance that you're using too small a tip or too low a temperature. Too low a temperature is easier to detect than too high, so work up to a temperature rather than down.

Use the Right Solder

Another factor in reducing the chances of heat damage is using the right solder. This means 63% tin and 37% lead (Sn63Pb37) with an activated rosin core (Fig. 5-2). Sn63Pb37 solder has the lowest melting point of the tin-lead alloys, and it is far lower than that of lead-free solders. Sn63Pb37 solder is suitable for soldering PC boards that use lead-free solder as well as leaded solder. An activated rosin core flux provides aggressive cleaning/wetting action to speed soldering.

TIP

Never use acid-core flux with electronics.

FIGURE 5-2 Kester 44 rosin-core solder 63/37, 0.031 inch. *(Courtesy of Kester.)*

WHAT'S SPECIAL ABOUT SN63PB37 SOLDER?

Many solder alloys are used to provide various desirable properties. Of all the alloys composed of only tin and lead, Sn63Pb37 solder is special. It is eutectic, which means that it melts and solidifies at a single temperature that's lower than the melting points of pure tin or pure lead or any other mixture of them (Fig. 5-3).

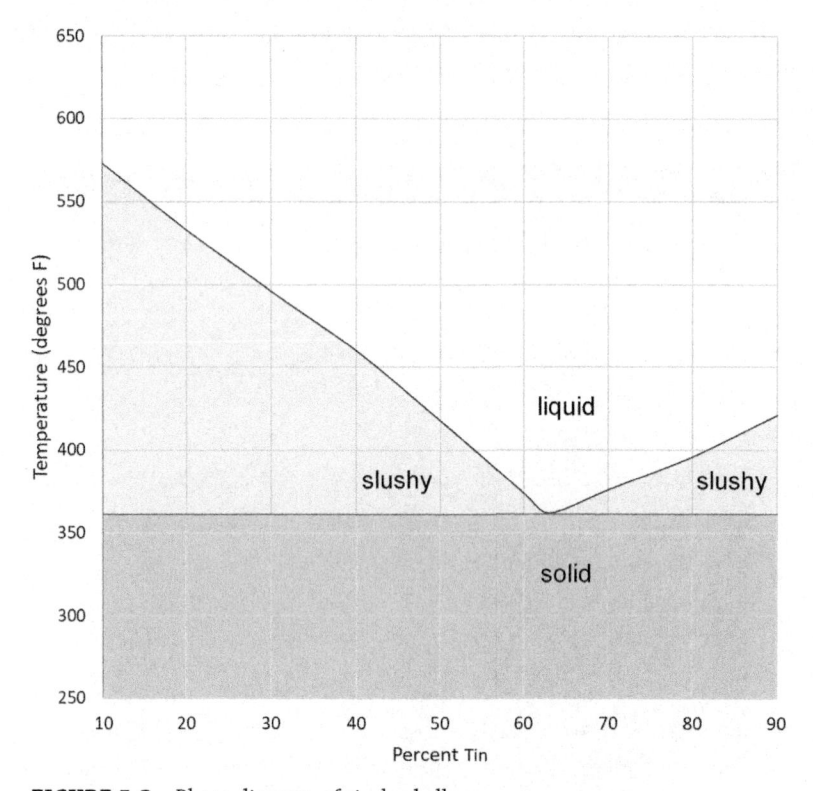

FIGURE 5-3 Phase diagram of tin-lead alloys.

For soldering, this lower temperature gives the advantage of reducing the likelihood of damaging electronic components during soldering. A second advantage is that it solidifies at a single temperature. The other alloys of tin and lead have a slushy phase as they cool from liquid to solid. If the connection is moved during this slushy phase, the result is a bad solder joint. Such joints have a dull appearance and are referred to as *cold solder joints*.

How to Solder

Follow these guidelines to produce quality solder joints (Fig. 5-4):

Before

- Use third hands or a vise to hold PC boards while you solder.
- Set your soldering iron to a medium temperature (600–700°F).
- Frequently clean your tip on a wet sponge and tin your tip with fresh solder.
- Make sure that items to be soldered are perfectly clean. Cleaning is best achieved with a Scotch-Brite pad.
- It's a good idea to pre-tin difficult-to-solder and large-surface-area joints such as stranded wire. Twist the strands together, and then heat and saturate them with solder.
- Soldering provides an excellent electrical bond but a weak mechanical bond. Make a good mechanical connection by twisting wires together or wrapping wires around a terminal.

During

- Use the side of the tip, the "sweet spot," not the very tip of the iron.
- Use the soldering iron to heat the joint, not the solder. A small amount of fresh solder on the tip will help conduct heat to the joint faster. The solder should be fed into and heated by the joint so that it will flow into and around the connection.
- Don't apply more solder than needed; excess can cause a short circuit.
- Pull the solder away first, then the iron. Smoothly slide the iron away to leave the joint neat.
- Avoid moving the joint or wire after removing the iron until the solder has solidified.

After

- A good solder joint should look like a volcano, not a ball or clump. It should be shiny and smooth, not dull.
- Wash your hands when you're done soldering.

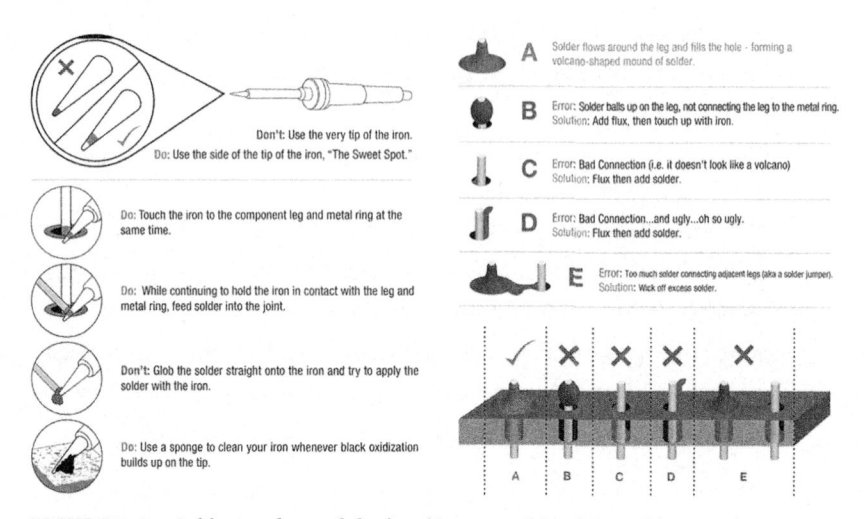

FIGURE 5-4 Soldering dos and don'ts. *(Courtesy of SparkFun Electronics.)*

Flux Removal after Soldering

The flux core of solder leaves a residue after soldering. Although some solder fluxes are called *no-clean*, Kester recommends removing flux residue after soldering. Solvents such as isopropyl alcohol can be used for this. Use an old toothbrush or cotton swabs dipped in solvent to gently clean your work.

Desoldering

When it comes to desoldering, you'll want to have an arsenal of tools and techniques to replace components without damaging the PC board. The electrical traces of many modern circuit boards are tiny and easily damaged by excess heat.

Solder Wick

Solder wick or solder braid (Fig. 5-5) is the soldering equivalent of a damp sponge for soaking up molten solder. It's embedded with flux that helps the solder flow into the braid. It comes in a range of widths to suit any job.

The technique is to position an unused section of the braid on top of the area to be desoldered, and then press a hot soldering iron against the braid to soak up molten solder (Fig. 5-6).

FIGURE 5-5   Chemtronics Soder-Wick.
(Courtesy of Chemtronics www.chemtronics.com.)

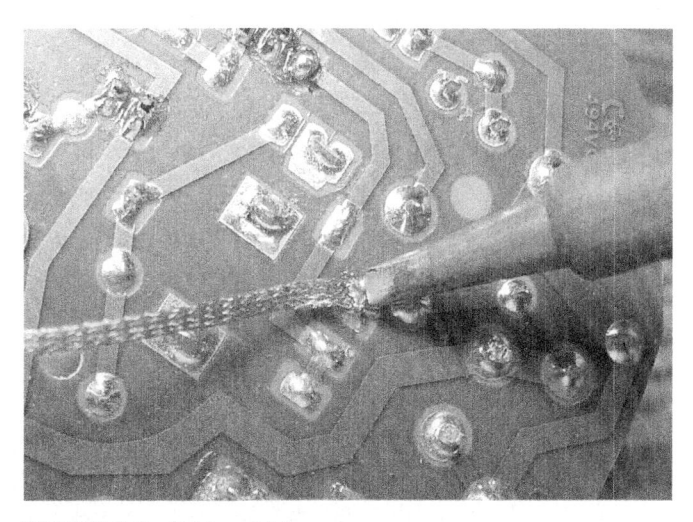

FIGURE 5-6   Solder wick in action.

Solder Suckers

Solder suckers or desoldering pumps (Fig. 5-7) are especially useful for removing large amounts of solder, but they can be used for any size job. The technique is to compress the spring-loaded plunger until it locks in place and then place the high-temperature tip next to the area to be desoldered (Fig. 5-8). Melt the solder with your soldering iron, and then press the trigger on the pump to draw a vacuum at the tip, which sucks in the molten solder. To clear solidified solder from the tip, press the spring-loaded plunger a little further than in the first step, and a rod inside will clear the tip. If the pump gets completely clogged, you'll need to remove the tip or barrel to clean out them out.

When buying a desoldering pump, choose an antistatic model. The vacuum pump action can create static electricity, and you don't want to zap components on your PC board when desoldering.

FIGURE 5-7 EDSYN Static-Safe Deluxe SOLDAPULLT DS017LS. *(Courtesy of EDSYN Inc.)*

Flux Pen

A flux pen (Fig. 5-9) provides a convenient way to dispense flux when and where you need it. It comes in handy for desoldering work, where normally no flux is present. The flux aids in heat transfer and promotes even solder flow, making desoldering easier and less likely to cause damage.

Add Solder

It might seem counterintuitive to add solder as the first step in desoldering, but there are two reasons to do so. If you don't have a flux pen, fresh solder is an easy way to add flux. Second, when you solder with the recommended Sn63Pb37 solder (see the section "Use the Right Solder"

FIGURE 5-8 Solder sucker in action.

FIGURE 5-9 Kester 186 Flux-Pen. *(Courtesy of Kester.)*

earlier in this chapter), the low-temperature alloy dilutes any higher-temperature alloys, such as lead-free solder, allowing better melting.

Use a Pin Vise

If you're having trouble fully desoldering a hole for a component lead and you're worried about damaging it with too much heat, consider using a pin vise (Fig. 5-10). The pin vise can hold a drill bit that's the same diameter as the component lead, and you can use it to drill out the solder with a few gentle twists. Be careful not to use a drill bit that's too large or you'll drill out a plated through-hole, causing more harm than good.

FIGURE 5-10 Pin vise in action.

Clip Instead of Desolder

Where PC board damage from desoldering is a big concern, consider not desoldering. This means clipping out the body of the defective component and soldering the replacement component to the pads or remaining leads of the old part. The following photos (Fig. 5-11) show the process of replacing a bad resistor.

When using this technique, keep in mind that soldering provides only modest mechanical strength, so use a wire tie, glue, or other method to mechanically anchor physically large components such as big electrolytic capacitors or power resistors. This may prevent them from ripping off the board if the product is later dropped on the floor.

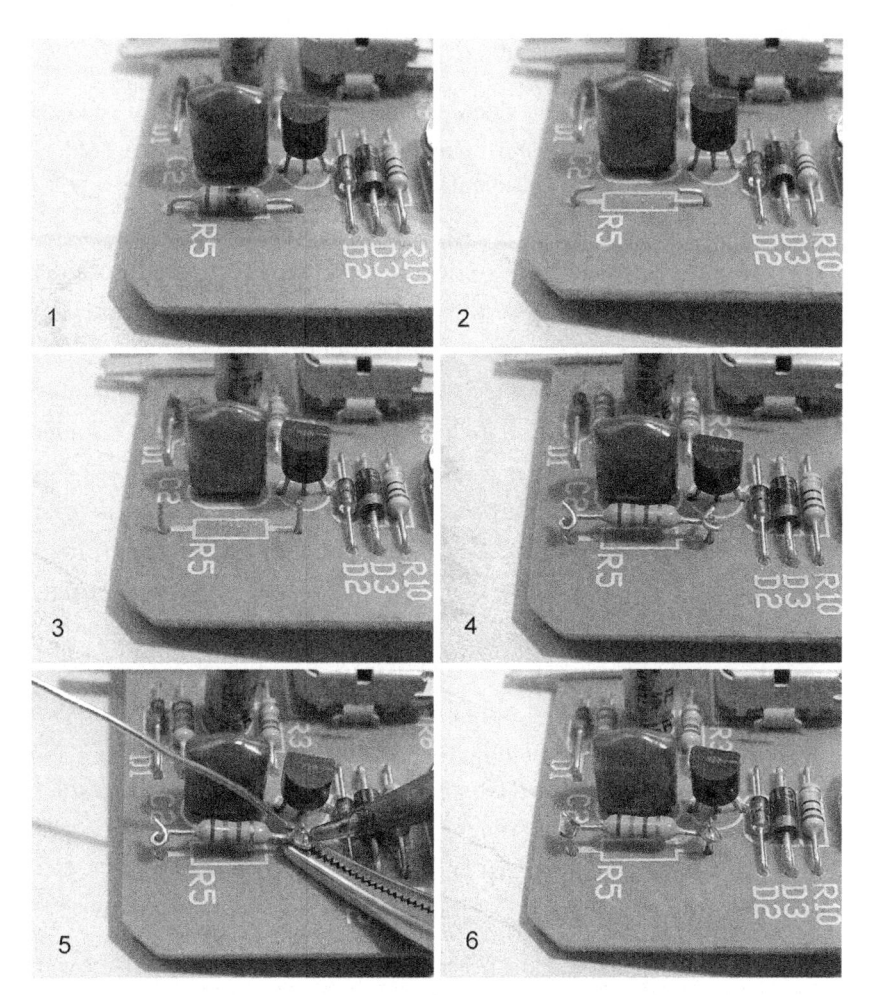

FIGURE 5-11 Clip instead of desolder: (1) bad resistor; (2) bad resistor with the body clipped out; (3) leads straightened; (4) new part in place (pre-tin the leads of the replacement part before putting in place, and twist the leads together if possible); (5) soldering in (try to minimize the time that you're heating the connection, and use a heat sink on the original leads if possible); (6) remove flux—repair complete.

PC Board Trace Repair

PC board traces can be easily damaged by high current or excessive heat. The excessive heat may come from a failed component or from desoldering. To repair a damaged trace, follow this procedure (Fig. 5-12):

1. Trim away the portions of the trace that have separated from the PC board using an X-Acto knife. Scrape off any charred areas of the board because these may be conductive (photo b).
2. Gently scrape the solder resist off the ends of the remaining intact trace to expose ¼ inch or so of bare copper. Tin with solder if desired (photo c).
3. Using solid pretinned wire, tack-solder the wire end to the first solder pad after the break or, better yet, the pad past that (photo d). Choose a wire gauge that can handle the current of the original trace (see Table 5-1).
4. Following the path of the original trace with the wire, tack-solder the wire at each pad along the way and at the exposed copper ends of the remaining intact trace (photo e).
5. Touch up each of the tack-soldered connections with more solder as needed (photo f). Remove flux.
6. If there's a risk of the wire shorting against something else, epoxy it down or use heat-shrink tubing on it.

If the trace was damaged by high current, be sure to repair the cause of the original problem or you'll be doing this again!

TABLE 5-1 Thinnest Wire for Repairing PCB Traces

PC Board Trace Width	Thinnest Wire for Repair
¹⁄₁₆ inch	26 AWG (American Wire Guage)
⅛ inch	24 AWG
¼ inch	20 AWG
½ inch	18 AWG

Soldering and Desoldering | 79

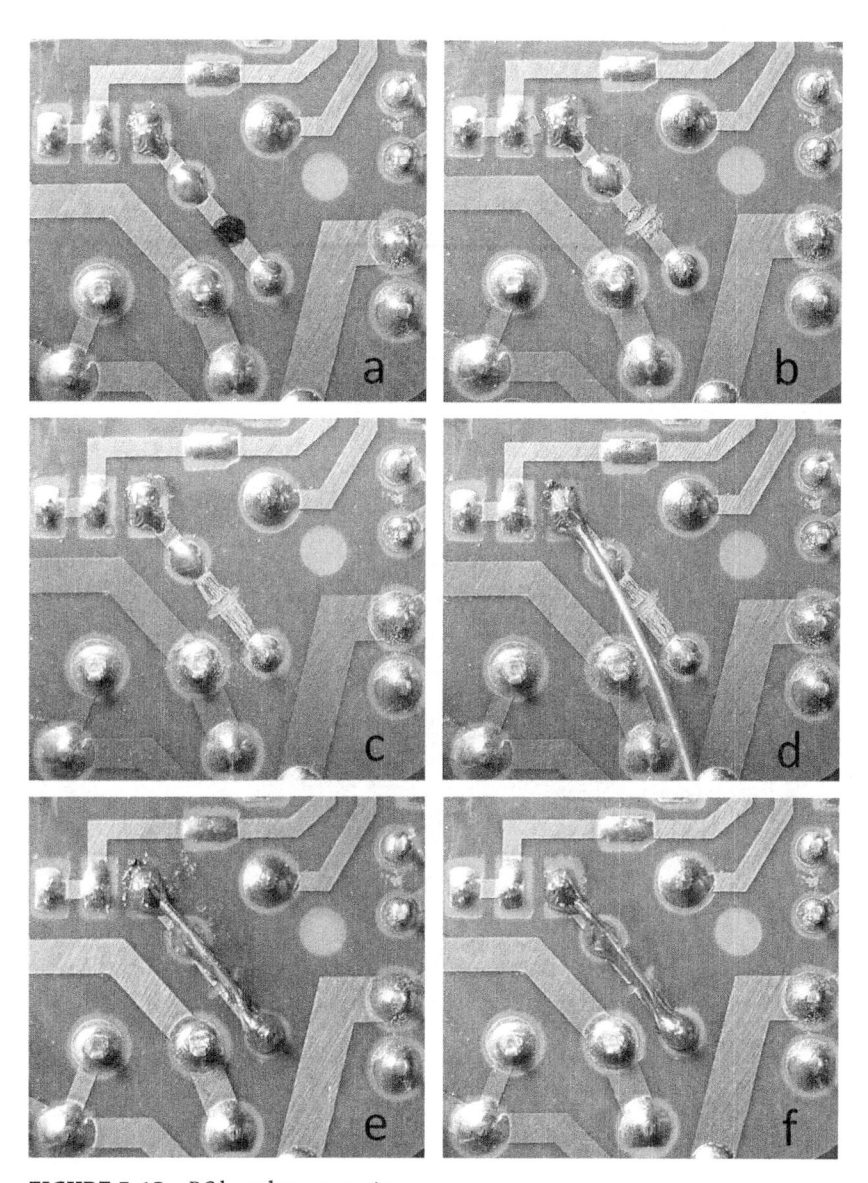

FIGURE 5-12 PC board trace repair.

Heat-Sinked Devices

Transistors, voltage regulators, and integrated circuits (ICs) often use metal heat sinks to help dissipate heat from the device into the surrounding air (Fig. 5-13). For a heat sink to do its job, there needs to be a good thermal connection between the component and the heat sink. When replacing heat-sinked devices, it is important to make sure that a good thermal connection is restored or the component will fail prematurely.

FIGURE 5-13 Heat-sinked devices.

Heat-Sink Compound

Heat-sink compound (also known as *thermal compound, thermal paste,* or *heat-sink grease*) is a white toothpaste-like substance used to fill the gaps between heat sinks and components (Fig. 5-14). Most heat-sink compounds are thermally conductive but electrically nonconductive.

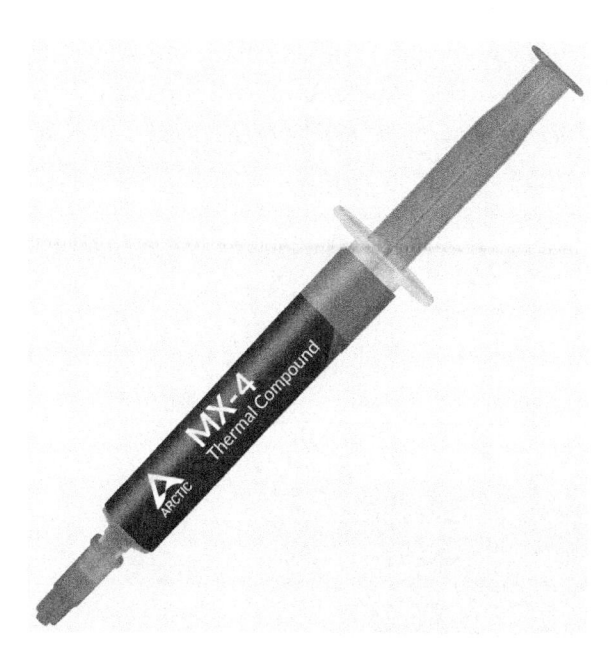

FIGURE 5-14 Arctic MX-4 thermal compound. *(Courtesy of Arctic www.arctic.ac.)*

"Liquid metal" compounds offer better thermal conductivity but are electrically conductive, so they can't be used where conductivity is a problem. Spillage also can cause problems; if it flows onto a circuit it can cause malfunctioning or damage.

When replacing parts that use heat-sink compound, always remove the old compound from all surfaces and replace it with fresh compound. Clean the old compound using a cotton swab with isopropyl alcohol. Heat-sink compound is messy and will stain your clothes and carpets, so be careful working with it.

Apply heat-sink compound sparingly. The idea is to fill the microscopic air gap with compound, nothing more. Heat-sink compound is a better thermal conductor than air but much worse than metal.

Thermal Pads

Thermal pads are a less messy alternative to heat-sink compound (Fig. 5-15). They come in sheet form and can be cut to fit almost any application. Like heat-sink compounds, thermal pads come in electrically nonconductive and conductive types. The electrically conductive pads provide better thermal performance, but they have the disadvantage of electrical conductivity.

FIGURE 5-15   Arctic thermal pad. *(Courtesy of Arctic www.arctic.ac.)*

The performance of pads is generally inferior to that of pastes, but the convenience of pads makes them a viable alternative. Expect to pay a premium for pads that outperform pastes.

COMPARING PADS AND PASTES

If you want to compare pastes and pads scientifically, compare their thermal conductivity specifications in watts per meter per kelvin (W/m-K). Higher numbers signify higher conductivity and are better. The thermal conductivity of a high-grade electrically nonconductive thermal paste is about 4 W/m-K. Electrically nonconductive thermal pads can have values ranging from less than 1 to more than 15 W/m-K.

Mica Insulators

Thin mica insulators and nonconductive washers (Fig. 5-16) are often used to prevent an electrical connection between the mating surfaces of components and heat sinks. These can be reused, but care must be taken to reinstall them in their original positions during repair.

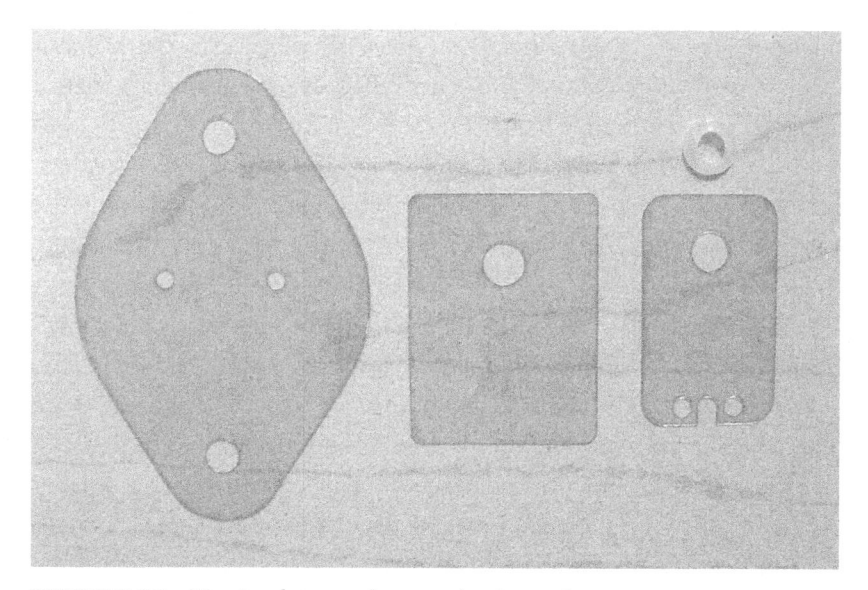

FIGURE 5-16 Mica insulators and nonconductive washers.

Bolt Down, Then Solder

Always bolt down the component before soldering. If you solder first and then bolt down the component, any misalignment of the parts during soldering will result in damage to the component or the PC board during the bolt down or poor contact between the heat sink and the component.

Soldering and Desoldering Surface-Mount Parts

The techniques for soldering and desoldering surface-mount parts are different from those for through-hole parts. During manufacturing, chip parts are glued to the PC board before being run through a solder bath, so removing them involves simultaneously breaking the glue bond and the solder bonds.

Chip resistors and capacitors are the easiest to desolder. Alternately heat each side of the component until it just wipes away with the soldering iron tip. Adding a little fresh solder to each side helps. Use solder wick to clean off the pads once the chip is removed.

Small outline integrated circuits (SOICs) are removed in essentially the same way. To remove them, you must add more solder such that it forms a glob across each row of pins. Then heat each side alternately until you can gently wipe the part off the board.

To solder on chip resistors and capacitors (Fig. 5-17), use tweezers with your left hand to position the chip part so that it is centered between the two pads. Using the soldering iron with your right hand, clean the tip on the damp sponge, and then get a small amount of fresh solder on the tip. Use this solder to tack the right side of the chip part to its pad. (If you're left-handed, reverse hands for this procedure.) Next, solder the left pad, and then touch up the solder on the right pad. If the chip isn't aligned properly, alternately remelt the solder on both sides, and the surface tension of the solder will help center the part. If you get too much solder on the pads, use solder wick to remove some.

Soldering SOICs is similar. Using tweezers, align the part on the pads, and tack down a single pin. Make sure that all the pins are aligned with the pads, and then solder each of the remaining pins. If solder bridges across any of the pins, this isn't a big deal. Just wipe away the excess solder with solder wick.

Soldering a quad flat pack (QFP) with hundreds of pins may seem intimidating, but it's not difficult. The strategy is to tack it into position, cover it with solder (ignoring shorts), and then remove excess solder using solder wick. The most critical step is tacking it into position so that all the pins are aligned with their pads. Start by tacking down one corner

Soldering and Desoldering | 85

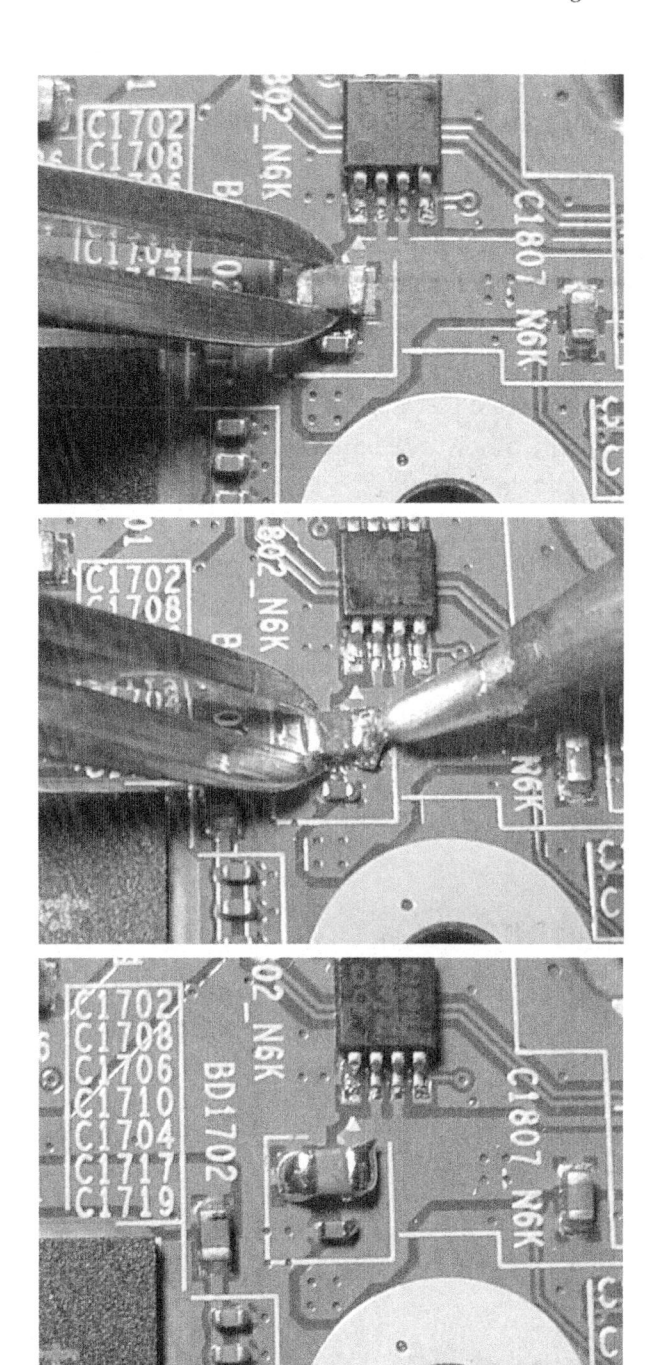

FIGURE 5-17 Soldering a chip part.

and then tacking down the opposite corner so that the part can't rotate. Use magnification to help with this step. Once the QFP is aligned, freely melt solder across all the pins. A flux pen can be used before soldering to improve solder flow. After soldering, use solder wick to remove the excess solder. Put the solder wick on top of the glob and gently press the soldering iron down on the wick. You will see the solder being sucked into the braid. The solder wick should be wiped in a direction parallel with the pins, not perpendicular to them because this will bend them. Repeat on all four sides of the QFP. Use magnification to check for shorts. If you find any, just wipe them away with heated solder wick.

CHAPTER 6

Using Test Equipment

This chapter explains the basics of how to use the two most important pieces of test equipment for electronics repair: digital multimeters and oscilloscopes. Refer to Chapter 4 for what to look for when buying.

This chapter isn't intended to replace the operating manual of any equipment. It's intended to show what the equipment can be used for and its ease of use.

Using a Digital Multimeter

Most digital multimeters provide measurement capabilities for DC and AC volts, resistance (ohms), and DC and AC current (amps). Additional features and operation vary by manufacturer and model. I'll be using the Fluke 115 multimeter (Fig. 6-1) as an example to describe the most commonly performed measurements when repairing electronics.

General Measurement Procedure

The general procedure for most measurements is

1. Turn on the meter. (The Fluke 115 automatically turns on by selecting a measurement type.)

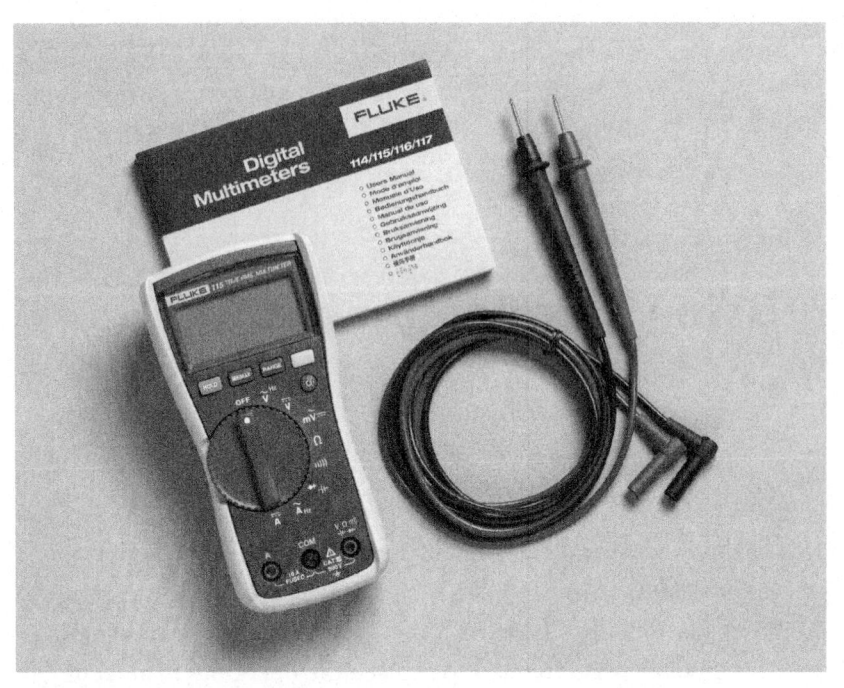

FIGURE 6-1 Fluke 115 multimeter.
(Reproduced with permission, Fluke Corporation.)

2. Select the measurement type (DC volts, AC volts, etc.) and measurement range. (On the Fluke 115, the knob, along with the yellow button, selects the measurement type. The 115 defaults to autorange. To enter the manual range mode, press the RANGE button repeatedly to select the desired range.)
3. Plug the probes into the multimeter jacks appropriate for the measurement. (For the Fluke 115, the black probe should be plugged into the COM jack and the red probe into the volt-ohm jack for all measurements except current.)
4. Touch the probe tips to the desired measurement points, and read the digital display on the meter.

Table 6-1 shows common examples of each measurement type.

TABLE 6-1 Examples of Multimeter Measurements

Measurement	Examples
DC volts	Checking whether an AC wall adapter is working
	Checking a battery
AC volts	Checking whether AC voltage is getting to an internal device point
	Checking whether a transformer is good
Resistance (ohms)	Checking whether a light bulb is good
	Measuring a resistor value
Continuity	Checking whether a fuse is good
	Checking whether a power cord is good
	Checking whether a switch is good
	Checking for shorts and opens on PC boards
DC current	Measuring how much DC current a device is using
AC current	Measuring how much AC current a device is using
Diode test	Checking whether a diode or transistor is good

CHECKING ALKALINE BATTERIES WITH A MULTIMETER

To check alkaline batteries with a multimeter, use the DC volts setting. The tricky part is knowing what voltage a good battery should be. Because you'll be measuring the battery without a load, the voltage should be higher than the nominal voltage. For example, fresh 1.5-volt batteries such as AAA, AA, C, and D cells should measure about 1.6 volts. If the voltage measures less than 1.2 volts, the battery should be replaced.

PIERCING PROBE TIP SUBSTITUTE

There are situations where you'd like to be able to probe a wire but don't have convenient access to either end. You can buy insulation-piercing probe tips, or you can use a sharp no. 11 X-Acto knife blade instead (Fig. 6-2).

FIGURE 6-2 X-Acto no. 11 blade as piercing probe tip.

Orient the blade in parallel with the wire so that the blade tip is most likely to slide between wire strands rather than cut into them. The hole in the blade is a convenient place to grip with your probe. Normally, a small cut in the insulation won't create any problems, but you can apply electrical tape over the cut if desired.

Continuity Mode

The continuity mode is a convenient method to check for opens and shorts. The meter beeps when low resistance is detected and is silent otherwise. (The Fluke 115 beeps when the resistance is less than 20 ohms.) The continuity mode is also useful for checking fuses, power cords, and switches.

Current Measurement

For routine repair work, current is less frequently measured than voltage or resistance. The measurement procedure is different, too. For one thing, you may need to plug your red probe into a different jack on the meter. (The Fluke 115 uses the A jack.) Second, the meter probes must be wired in series with the circuit to be measured. See Figure 6-3.

Never place the probes in parallel with a circuit or component when a meter is in the current measurement mode or you will short out the circuit, possibly causing damage.

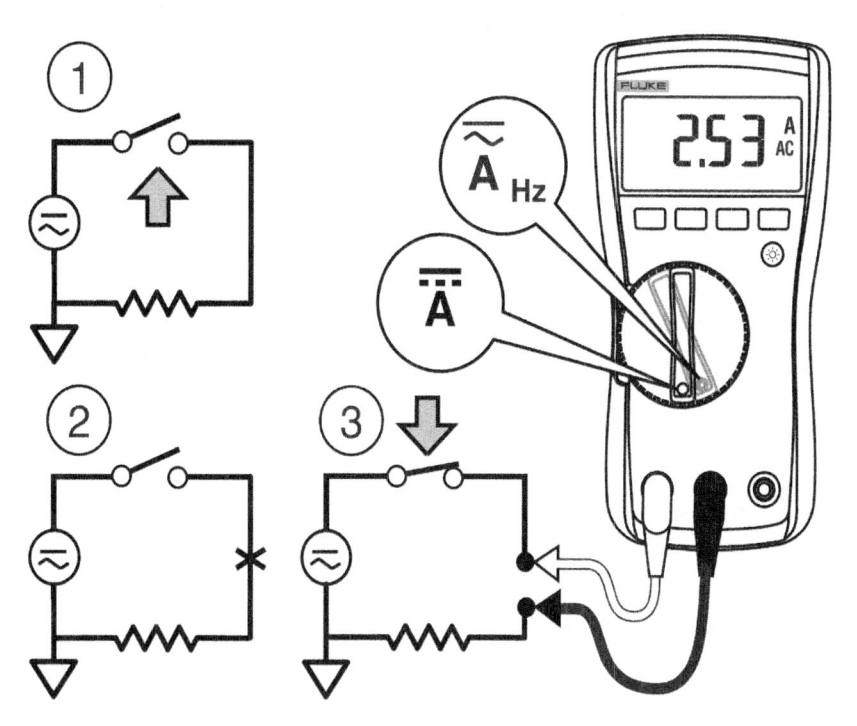

FIGURE 6-3 Current measurement with a multimeter. *(Reproduced with permission, Fluke Corporation.)*

Diode Test

Diode test is probably the least understood feature on a multimeter, but it's indispensable for determining whether diodes or transistors are bad.

In the diode test mode, the meter applies a DC current and measures the resulting voltage. If you touch the two probes together, you'll see that the voltage is about zero. This means current is flowing. If the two probes are not touching anything, the meter will show an OL (Over Limit) condition. This means current is not flowing.

Testing diodes is easy (Fig 6-4). A diode is a one-way valve for electric current. If you connect the probes to a good diode, the meter will show about 0.6 volts in one direction and OL with the probes reversed. For most silicon diodes, 0.6 volts is normal and is called the *diode drop*. The diode drop for other types of diodes varies. For Schottky diodes, often used in power supply circuits, it can be as low as 0.2 volts. The important thing is that you see a voltage between 0.2 and 0.7 volts with one polarity and an OL condition with the reverse polarity. If you measure OL with both polarities or a voltage of less than 0.2 volts for both polarities, the diode is bad.

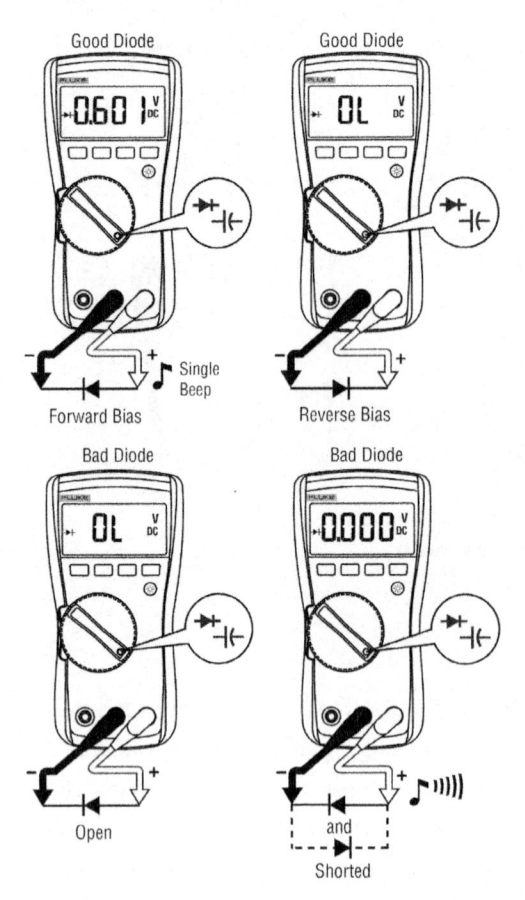

FIGURE 6-4 Diode test with a multimeter. *(Reproduced with permission, Fluke Corporation.)*

Using an Oscilloscope

Learning to use an oscilloscope is a lot like learning to drive a car. All the controls might seem overwhelming at first, but it doesn't take long to get familiar with the basic operation. This section will show the process for compensating a scope probe. Along the way, all the basic scope controls will be used. The point is to show what it's like to drive a scope. I'll be using the Tektronix TBS1032B (Fig. 6-5) as an example for the process.

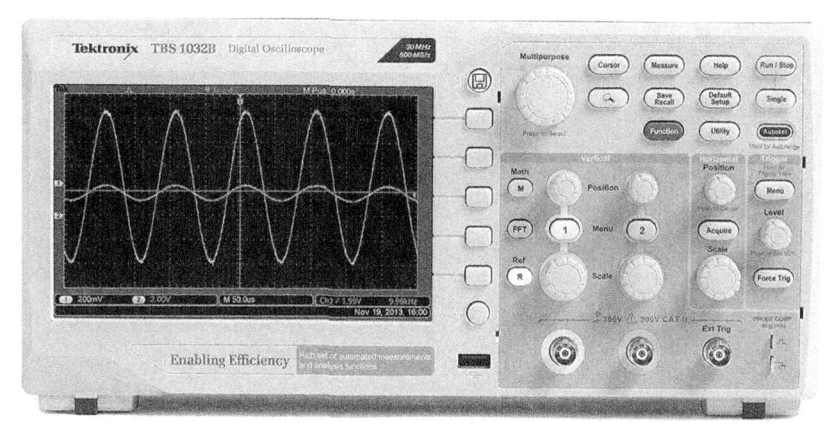

FIGURE 6-5 Tektronix TBS1032B two-channel 30-MHz digital storage oscilloscope. *(Copyright © Tektronix. Reprinted with permission. All rights reserved.)*

Scope Probes

Oscilloscope probes (Fig. 6-6) are the most common way to connect an oscilloscope to a circuit. These have a spring-loaded hooked tip for hooking onto a pin, wire, or component lead. The hooked tip can be pulled off to expose a pointed needle tip. An alligator clip is used for the ground connection. Don't connect the alligator clip to anything but ground or you may damage your circuit.

Most scope probes are 1X/10X. This means that there's a switch on them that lets you select between 1X operation and 10X operation. The term 10X is misleading because it sounds like it would be providing amplification by a factor of 10. In fact, it is just the opposite: 10X *divides* the signal by 10. The 10X mode is the best mode to use for almost every

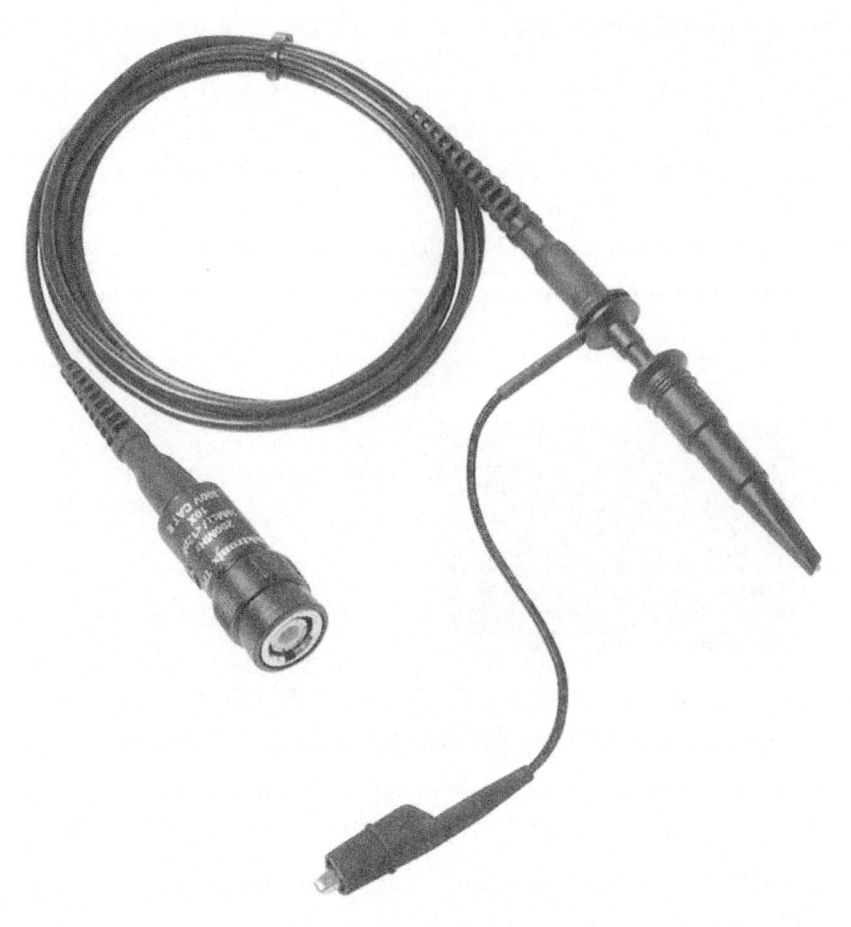

FIGURE 6-6 Tektronix TPP0051 1X/10X oscilloscope probe.
(Copyright © Tektronix. Reprinted with permission. All rights reserved.)

measurement because it reduces the loading of the circuit by the probe. The 1X mode is useful is for very small signals where dividing by 10 would make them too small to display.

Probe Compensation

All 10X probes have an adjustment screw on them for high-frequency compensation. This adjusts the probe to compensate for the input capacitance of the oscilloscope. Compensation should be performed whenever you use a probe on a different scope. Oscilloscopes usually provide a square-wave test signal that you can use for compensation.

Probe Compensation Procedure

1. Turn the oscilloscope on.
2. Connect the probe to the Channel 1 input on the scope. To do this, align the slot in the probe connector with the key on the Channel 1 BNC connector, push to connect, and twist clockwise to lock the probe in place.
3. Set the switch on the probe to the 10X position.
4. Connect the probe tip to the PROBE COMP ~5V@1kHz terminal of the oscilloscope and the alligator clip of the ground lead to the PROBE COMP terminal.
5. Set the Channel 1 input to the 10X mode. (For the TBS1032B, push the Channel 1 Menu button. Select Probe > Voltage > Attenuation > 10X.)
6. Set the Channel 1 input to DC coupling. (For the TBS1032B, push the Channel 1 Menu button. Select Coupling > DC.)
7. Set the Channel 1 vertical scale to 1 volt/division. (For the TBS1032B, rotate the Channel 1 Vertical Scale knob as necessary. The display shows the value.)
8. Use the Channel 1 vertical position control to center the waveform in the display.
9. Set the horizontal scale to 500 μs/division. (For the TBS1032B, rotate the Horizontal Scale knob as necessary. The display shows the value.)
10. Set the scope to trigger on Channel 1. Select edge triggering, rising-edge slope, auto triggering mode, AC-coupled triggering. (For the TBS1032B, push the Trigger Menu button. Turn the Multipurpose knob to highlight the desired options from the pop-out menu. Push the knob to select the choice.)
11. Adjust the trigger level to the 50% position of the waveform or as needed to produce a stable display.

12. Adjust the screw on the scope probe to produce a square wave without overshoot or undershoot. See Figure 6-7 for guidance.

Overcompensated

Undercompensated

Compensated correctly

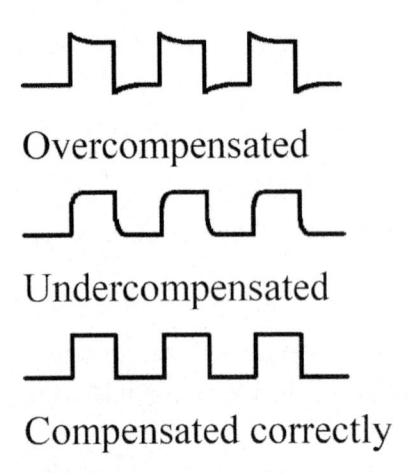

FIGURE 6-7 Scope compensation waveforms.
(Copyright © Tektronix. Reprinted with permission. All rights reserved.)

WHAT IS TRIGGERING?

The three main oscilloscope controls are Horizontal Scale, Vertical Scale, and Trigger Level. The first two are easy to understand, but Trigger Level is less clear.

Triggering works like a strobe light for electrical signals. A strobe light can be used to visually freeze the motion of a spinning wheel by synchronizing the flashing of the light with the rotation of the wheel. The triggering system on an oscilloscope looks for a recurring feature in the input signal and uses it to synchronize the display on the screen.

The triggering system of most oscilloscopes provides several settings to help stabilize the display to the signal feature you want. The Trigger Level control adjusts the voltage threshold of the trigger. Other triggering settings include positive or negative edge selection, AC or DC coupling, and filtering.

CHAPTER 7

Part Identification, Testing, and Substitution

This chapter is about how to identify parts, tell whether they are working, and find a substitute, if needed. Unless stated otherwise, the procedures for checking parts in this chapter assume that the part is electrically out of circuit. This means that it is okay to have one leg of a component still soldered in. Chapter 8 covers in-circuit testing of components in detail.

Counterfeit and Off-Brand Parts

It's tempting to go with the cheapest price when shopping for replacement parts. But not all parts are created equal. Counterfeiting of semiconductors is surprisingly common. Experts have estimated that as many as 15 percent of all spare and replacement integrated circuits purchased by the Pentagon are counterfeit, and the military buys from reputable suppliers. E-waste is the source of many counterfeit and re-marked semiconductors. As with many counterfeit goods, a disproportionate number of fake chips are traced back to China.

Off-brand parts are often designed to look like their name-brand counterparts but don't have the same quality. Electrolytic and tantalum capacitors in particular should only be purchased from name-brand manufacturers.

To avoid problems down the road, use reputable suppliers when buying replacement parts, especially semiconductors and capacitors. Use name-brand parts whenever possible.

Resistors

Resistors (Fig. 7-1) are the most common component on most circuit boards and may be through-hole or surface mount. Their resistance can be checked using the resistance mode of a multimeter. Resistors usually fail as an open circuit and may show visible signs of overheating such as darkening, burning, or deformation.

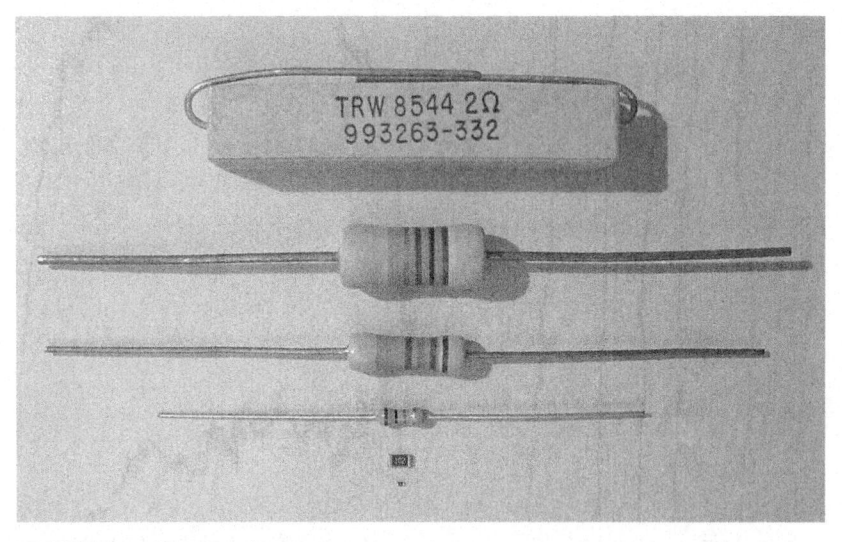

FIGURE 7-1 Variety of resistors.

In electronics, most resistors are specified in ohms (Ω), or in kilohms (kΩ), which are one thousand ohms. Large value resistors may be specified in megohms (MΩ), which are one million ohms, or gigohms (GΩ), which are one billion ohms.

Identifying Resistance Value and Tolerance

Most resistors with leads are easily identified by the bands of color used to show their nominal resistance value and tolerance. The color code for four-band resistors is shown in Table 7-1.

TABLE 7-1 Resistor Color Code

Band 1 = First Digit	Band 2 = Second Digit	Band 3 = Multiplier	Band 4 = Tolerance
Black = 0	Black = 0	Black = ×1	—
Brown = 1	Brown = 1	Brown = ×10	—
Red = 2	Red = 2	Red = ×100	—
Orange = 3	Orange = 3	Orange = ×1,000	—
Yellow = 4	Yellow = 4	Yellow = ×10,000	—
Green = 5	Green = 5	Green = ×100,000	—
Blue = 6	Blue = 6	Blue = ×1 million	—
Violet = 7	Violet = 7	Violet = ×10 million	—
Gray = 8	Gray = 8	Gray = ×100 million	—
White = 9	White = 9	White = ×1 billion	—
—	—	Gold = ×0.1	Gold = 5%
—	—	Silver = ×0.01	Silver = 10%
—	—	—	None = 20%

For example, a resistor with color bands brown/black/red/gold would be 10 × 100 ohms = 1,000 ohms, with a 5% tolerance. Because gold and silver are never used for the first digit, you know which direction to read the color bands.

For 1 percent resistors, a band representing the third digit is inserted between the second digit and the multiplier. The last band would be brown to indicate 1 percent.

Surface-Mount Device (SMD) Marking Codes

Surface-mount chip resistors don't have color bands; instead, the resistance value is printed on them using a numeric code. Very small chip resistors rarely have any markings—there just isn't enough space.

Three marking systems are in common use: the three-digit system, the four-digit system, and the EIA-96 system. The three- and four-digit systems work like the color-coding system for resistors with leads. The first two or three digits show the value, and the last digit is the multiplier. For example:

$$100 = 10 \times 10^0 \text{ ohms} = 10 \text{ ohms} (\textit{not } 100 \text{ ohms!})$$

$$4991 = 499 \times 10^1 \text{ ohms} = 4{,}990 \text{ ohms}$$

The letter "R" is used to indicate the position of a decimal point for resistance values lower than 10 ohms. The tolerance isn't indicated with the three- and four-digit systems.

The EIA-96 code is used for chip resistors with 1 percent tolerance. It uses a two-digit number to indicate the value (Table 7-2) followed by a letter to indicate the multiplier (Table 7-3). For example:

$$68H = 499 \times 10 \text{ ohms} = 4{,}990 \text{ ohms}, 1 \text{ percent}$$

In addition to the three systems described, manufacturer-specific marking systems are also used. If the code doesn't jive with an ohmmeter reading, an alternate code may be the reason.

TABLE 7-2 EIA-96 Chip Resistor Number Code

Code	Value	Code	Value	Code	Value	Code	Value	Code	Value	Code	Value
01	100	17	147	33	215	49	316	65	464	81	681
02	102	18	150	34	221	50	324	66	475	82	698
03	105	19	154	35	226	51	332	67	487	83	715
04	107	20	158	36	232	52	340	68	499	84	732
05	110	21	162	37	237	53	348	69	511	85	750
06	113	22	165	38	243	54	357	70	523	86	768
07	115	23	169	39	249	55	365	71	536	87	787
08	118	24	174	40	255	56	374	72	549	88	806
09	121	25	178	41	261	57	383	73	562	89	825
10	124	26	182	42	267	58	392	74	576	90	845
11	127	27	187	43	274	59	402	75	590	91	866
12	130	28	191	44	280	60	412	76	604	92	887
13	133	29	196	45	287	61	422	77	619	93	909
14	137	30	200	46	294	62	432	78	634	94	931
15	140	31	205	47	301	63	442	79	649	95	953
16	143	32	210	48	309	64	453	80	665	96	976

TABLE 7-3 EIA-96 Chip Resistor Letter Code

Code	Multiplier
Z	0.001
Y or R	0.01
X or S	0.1
A	1
B or H	10
C	100
D	1,000
E	10,000
F	100,000

Power Rating

Another parameter that's important when replacing a resistor is the power rating. The power rating of a replacement resistor should always be equal to or greater than that of the original, or you risk burning it out. If the part isn't marked and you don't have access to service data showing the wattage, you'll have to guess based on the physical size of the original. Table 7-4 shows typical dimensions for through-hole resistors of various power ratings. When in doubt, go with a higher power rating.

TABLE 7-4 Resistor Size Versus Power Rating

Power rating (watts)	Body length (inch/millimeter)	Body diameter (inch/millimeter)
⅛	0.12/3	0.07/1.8
¼	0.26/6.5	0.10/2.5
½	0.33/8.5	0.13/3.2
1	0.43/11	0.18/4.5
2	0.60/15	0.20/5

Substitution

Resistors of every kind are readily available from the major electronics suppliers, so substitution isn't usually needed for replacement. But if you want to make do with parts you have on hand, you can use series or parallel resistors to create a desired value. Use the formulas in Chapter 3 to calculate the resistance. The power handling will be the sum of the individual power handlings when all the series or parallel resistors are the same value. If they are of different values, the calculation gets more complicated. For example, a 1-kilohm, 1-watt resistor in parallel with a 10-kilohm, 1-watt resistor would have a combined power rating of 1.1 watts, not 2 watts.

FIGURE 7-2 Variety of potentiometers.

Potentiometers

Potentiometers (Fig. 7-2), also called pots, are recognizable by a knob, screwdriver slot, or slider for adjustment and three terminals. Adjustable capacitors can look similar but are used less frequently and usually have only two terminals. Because pots are mechanical devices, they're a common source of problems.

Potentiometers usually fail open as a result of overheating or mechanical stress. It's also common for the wiper contact to oxidize, creating a bad connection.

The resistance value of potentiometers is almost always stamped on the body. To test a potentiometer out of circuit, first measure the resistance between the two outer terminals. It should match the value shown on the body. Next, measure the value between one of the outer terminals and the center terminal as you rotate the control over its full range. It should change smoothly and continuously between 0 ohms and the nominal resistance value.

In circuits, pots are often used as variable resistors, and the center wiper terminal is connected to one of the outer terminals with a wire or circuit board trace. The two most common applications for pots are as user controls and for calibration.

Pots Used as Controls

If a potentiometer functions as a user-operated control and is varied during normal product operation, check it in-circuit by rotating the control over its full range. The resistance from an outer terminal to the center terminal should change smoothly and continuously between 0 ohms and some maximum resistance value. Make sure that the outer terminal you use for the measurement isn't one that's electrically connected to the wiper. The maximum resistance value will likely be less than the nominal resistance value of the pot because of circuit loading.

It's common for the wiper to become oxidized, making poor contact with the resistive substrate in the pot. Contact sprays often can fix this problem. CAIG DeoxIT Fader F5 is specially made for pots and works well if you can find an access point for the spray into the pot. If you're lucky, there's a vent hole in the pot cover where you can direct the spray tube. Sometimes the pot cover needs to be removed to get the contact cleaner where it needs to go. Other times there's no way to get contact cleaner inside, and the noisy pot must be replaced.

Pots Used for Calibration

I've seen cases where the only thing wrong was an oxidized wiper contact on a calibration pot. If a calibration pot is suspect, carefully mark the setting of the pot with a fine-tipped Sharpie marker. Make sure that the power is off during this process. It's also a good idea to measure and write down the resistance value between the wiper and an outer terminal, especially for multiturn pots. Now move the control back and forth 10 or more times to remove the layer of oxidation from the wiper. Return the control to its original setting using the Sharpie marks or resistance measurement. If you have the service manual, use this opportunity to properly set the calibration pot.

Other Pot Parameters and Substitution

Besides resistance value, important parameters to consider when replacing a pot are the power rating, number of turns, taper, shaft type, number of gangs, and built-in on/off switch. That's a lot to worry about and can make replacement challenging. When an exact replacement isn't available, shaft type is the first parameter to consider changing. You can buy shaft adapters to convert split or spline shafts to solid shafts or smaller to larger shafts.

It's advisable to use a replacement pot with a power rating equal to or greater than that of the original. If unknown, matching physical size is your best bet.

Most pots are single turn, but where precision is required, multiturn pots are often used. This is easily checked by counting the number of rotations of the control.

Taper refers to the resistance versus setting characteristic of the pot. Almost all pots are linear taper, which means that you get 50 percent of the resistance at the 50 percent setting of the pot. Audio taper is often used for volume controls in audio equipment to provide a more pleasing control characteristic. Other tapers also exist for specialty applications. The way to check for this is by measuring the wiper terminal to outer terminal resistance at the 50 percent position. If it's close to 50 percent resistance, you most likely have linear taper.

Ganged pots allow one knob to simultaneously control multiple pots. They are most commonly used for stereo volume controls. Ganged pots can be sold preassembled with the number of gangs you need or as a style where you assemble the sections yourself.

Concentric pots are used in some older car stereos and test equipment. These are two potentiometers individually adjusted through concentric shafts.

Capacitors

Capacitors (Fig. 7-3), also called caps, come in a variety of types, including electrolytic, tantalum, film, and ceramic. Capacitance is measured in farads, but one farad is so large that the only place you'll ever see a capacitor that big is at a car stereo competition. For electronic repair, most caps are specified in microfarads, abbreviated µF, which are one-millionth of a farad. Very small values are specified in picofarads, abbreviated pF, one-trillionth of a farad. Sometimes nanofarads, abbreviated nF, one-billionth of a farad, are specified, but this is unnecessary because 1 nanofarad equals 0.001 microfarads or 1,000 picofarads.

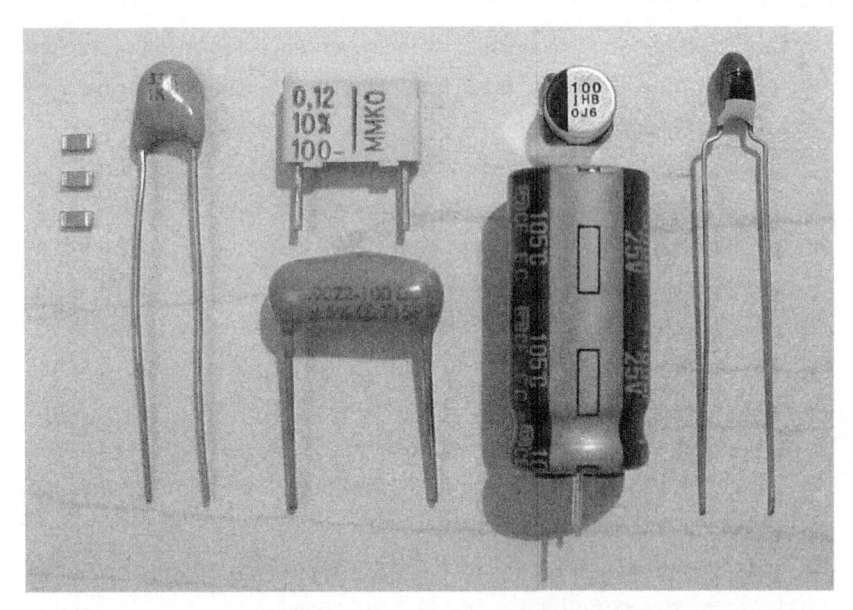

FIGURE 7-3 Variety of capacitors: chip, ceramic, film, electrolytic, and tantalum.

Electrolytic Capacitors

Electrolytic capacitors look like metal cans and range in size from smaller than a pencil eraser to larger than a soup can. Through-hole parts may have axial or radial leads (see Figure 7-4). Values cover the range from 0.1 µF to more than a farad. Electrolytics are usually polarized, which means that they have plus and minus terminals and can only be used

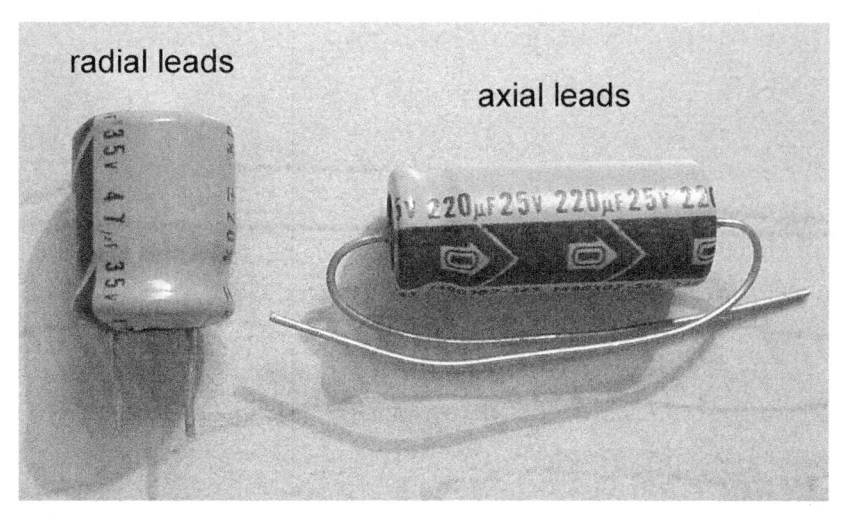

FIGURE 7-4 Radial versus axial leads.

for DC voltage. Nonpolarized electrolytics also exist; their main use is for crossover filters inside loudspeaker cabinets. Electrolytics generally provide the highest capacitance in the smallest size and at the lowest cost. This is where their advantages end. They are not very useful at high frequencies, and with the proliferation of low-quality parts in modern electronics, electrolytics are probably the least reliable electronic component of any. If you've got a broken device, chances are good that the problem is one or more bad electrolytic capacitors.

Electrolytic caps usually degrade gradually, with the capacitance value decreasing and the equivalent series resistance (ESR) increasing over time. At some point, the ESR can get so high that internal heating causes the part to fail catastrophically. This causes it to bulge or split open at the top or leak electrolyte. It's easy to spot these just by looking.

Other times an electrolytic cap may fail with no visible evidence and must be measured. You can measure the capacitance with a capacitance meter (some multimeters have this capability). A more accurate method is to measure the ESR with an ESR meter. Another advantage of an ESR meter over an ordinary capacitance meter is that it allows in-circuit testing so you don't need to remove caps from the circuit board. See Chapter 4 for more information.

Discharging Electrolytic Capacitors

Electrolytic capacitors can store high voltages for minutes or even hours. Large-capacity high-voltage electrolytics are common in the power supply circuits of many products and pose a potential hazard. You can use a DC voltmeter to check for electrolytics with a stored charge. When working around them, it's a good idea to discharge any large capacitors measuring more than a few volts. The residual voltage can affect in-circuit measurements even if it's well below hazardous levels.

Some people use a screwdriver to electrically short the two terminals of a capacitor to discharge it, but this is a hazardous practice. It can produce a big spark and vaporize a chunk out of your screwdriver tip or capacitor terminal. In some circumstances, the capacitor can even explode. It's better to discharge a capacitor through a power resistor (Fig. 7-5).

A 47-ohm, 100-watt resistor is a good choice for many situations. It will safely discharge a 2,200-µF or smaller capacitor charged to 200 volts or less or a 4,700-µF or smaller capacitor charged to 50 volts or less. See the sidebar on the science of discharging capacitors for more information.

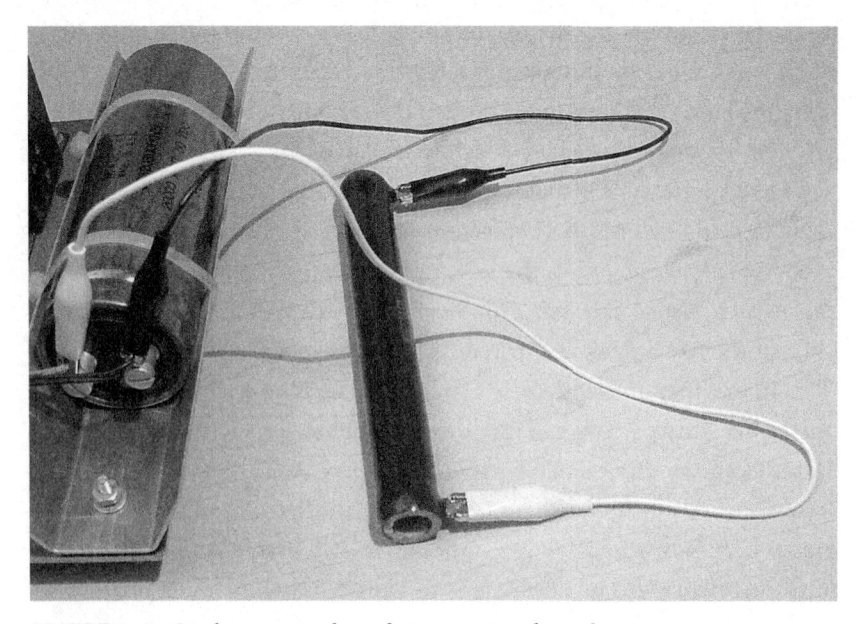

FIGURE 7-5 Discharging an electrolytic capacitor through a resistor.

THE SCIENCE OF DISCHARGING CAPACITORS

If you're wondering about the best resistor to use for discharging a capacitor, it's possible to calculate the ideal resistance value and power rating for discharging any capacitor. If the resistance value is too high, it will take too long to discharge the cap. Nobody wants to wait 10 seconds. If the resistance value is too low, the power rating of the resistor will need to be unnecessarily high to handle the initial surge of discharge current.

Table 7-5 shows ideal discharging resistors for a range of capacitors. The table assumes that the capacitor could have up to the rated voltage on it at the time of discharge. It also assumes that you want to discharge the capacitor voltage down to 1 volt and that you want to do it in less than 1 second. The final assumption is that the power resistor can safely handle 10 times its rated power for a fraction of a second.

TABLE 7-5 Ideal Discharging Resistors for Capacitors

Capacitance (µF)	Volts DC	Resistance (ohms)	Power (watts)
100	50	2556	0.1
220	50	1162	0.2
470	50	544	0.5
1,000	50	256	1.0
2,200	50	116	2.2
4,700	50	54	4.6
10,000	50	26	9.8
100	100	2171	0.5
220	100	987	1.0
470	100	462	2.2
1,000	100	217	4.6
2,200	100	99	10.1
4,700	100	46	21.6
10,000	100	22	46.1

(continued on next page)

TABLE 7-5 Ideal Discharging Resistors for Capacitors (*continued*)

Capacitance (µF)	Volts DC	Resistance (ohms)	Power (watts)
47	200	4016	1.0
100	200	1887	2.1
220	200	858	4.7
470	200	402	10.0
1,000	200	189	21.2
2,200	200	86	46.6
4,700	200	40	99.6
22	400	7587	2.1
47	400	3551	4.5
100	400	1669	9.6
220	400	759	21.1
470	400	355	45.1
1,000	400	167	95.9
2,200	400	76	210.9

If you decide to use this table, exactness is not required. You can use a higher resistance, but you'll need to discharge the capacitor longer.

A special case that falls well outside this table is microwave oven capacitors. These are usually about 1 µF and are rated at about 2,000 volts. Use a 100-kilohm, 10-watt resistor to discharge them. Use extra caution when working with the possibility of very high voltages such as this.

Replacing Electrolytic Capacitors

When replacing electrolytics, parameters to consider include capacitance value, voltage, physical size, lead configuration, ESR, and temperature rating. When ordering parts, choose the same capacitance value and voltage rating as the original part. It's generally not a problem to substitute electrolytics with a slightly higher capacitance value or voltage rating.

If you substitute radial for axial parts (or vice versa), slide some heat-shrink tubing on the exposed leads to reduce the chances of an accidental short if something should touch the bare wires.

For electrolytics in switch-mode power supplies, always use low-ESR caps. Always order electrolytics with a 105°C temperature rating—they have better reliability. Be sure to order only name-brand parts from a reputable supplier. Name-brand electrolytic capacitor manufacturers include Cornell Dubilier, Nichicon, Panasonic, Rubycon, TDK, United Chemi-Con, and Vishay.

Observe the polarity markings on the capacitors and PC boards to make sure that you don't put a part in backwards. The negative terminals are usually indicated with a stripe or shaded area. Finally, consider proactively replacing other large-value caps in the circuit you're working on; they may be ready to fail.

WHAT'S SPECIAL ABOUT 105°C?

105°C is hotter than the boiling point of water. It's unlikely that your electrolytic capacitors will be used at or near this temperature, but you should always buy electrolytics rated at 105°C. The reason is that the temperature rating is an accurate indicator of electrolytic lifespan.

Within limits, the life of a capacitor doubles for every 10°Celsius decrease in operating temperature. Equivalently, the life of a capacitor doubles for every 10°Celsius increase in temperature rating. Caps rated for 105°C can be expected to last about four times longer than equivalent caps rated at 85°C.

Tantalum Capacitors

Solid tantalum through-hole capacitors are pearl or teardrop shaped and are polarized. Surface-mount versions are common and can be identified because they don't look like a can, and they have a polarity indication. Tantalum caps usually have their positive terminals marked with a plus (+) sign. Because electrolytics have their negative terminals marked with a minus (−) sign, this can cause dangerous confusion.

Typical values cover the range from 0.1 to 1,000 µF. Their capacitance and voltage values may be printed on them or color-coded. Table 7-6 shows the color code for through-hole parts.

TABLE 7-6 Tantalum Capacitor Color Code

Top = First Digit	Middle = Second Digit	Dot = Multiplier	Bottom = Voltage
Black = 0	Black = 0	Black = ×1 µF	Black = 10 volts
Brown = 1	Brown = 1	Brown = ×10 µF	—
Red = 2	Red = 2	Red = ×100 µF	—
Orange = 3	Orange = 3	—	—
Yellow = 4	Yellow = 4	—	Yellow = 6.3 volts
Green = 5	Green = 5	—	Green = 16 volts
Blue = 6	Blue = 6	—	Blue = 20 volts
Violet = 7	Violet = 7	—	—
Gray = 8	Gray = 8	Gray = ×0.01 µF	Gray = 25 volts
White = 9	White = 9	White = ×0.1 µF	White = 3 volts
—	—	—	Pink = 35 volts

Tantalum caps have much better high-frequency performance than electrolytics and don't degrade over time, but they are extremely sensitive to overvoltage or reverse voltage conditions, which can destroy them. Tantalums tend to fail at turn-on. They almost always fail as a short circuit. Combined with the fact that tantalums are usually used for power supply bypassing, this means that if a product fails at turn-on with symptoms that a power supply is down, a tantalum cap is a prime suspect. *Bypassing* is an electronics term that means using a capacitor connected between a point in a circuit (normally a DC power supply) and ground to "bypass" high-frequency noise.

Film Capacitors

Film capacitors are capacitors that use dialectics made of polypropylene, polyester, polyethylene naphthalate, or polyphenylene sulfide. Film capacitors are nonpolarized and come in through-hole and surface-mount styles. Values normally cover the range from picofarads to hundreds of microfarads.

Film capacitors have excellent high-frequency performance and low ESR and are very stable over temperature ranges. Some have self-healing properties when subjected to surges. In short, they come close to being ideal capacitors.

Although highly reliable, film caps can fail. Film caps can fail open, shorted, or degraded. When ordering replacement parts, choose the same capacitance value and equal or higher voltage rating as the original.

Ceramic Capacitors

Ceramic capacitors are the most produced and used capacitors in electronics equipment, with annual quantities in the trillions. They are nonpolarized and come in surface-mount and through-hole styles. Values normally cover the range from fractions of picofarads to hundreds of microfarads.

Ceramic caps provide good high-frequency performance and low ESR. They are divided into Class 1 and Class 2 types. Class 1 is intended for filters or resonant circuits, and Class 2 is intended for power supply bypassing. Table 7-7 shows some of the key differences.

TABLE 7-7 Class 1 Versus Class 2 Ceramic Capacitors

Parameter	Class 1	Class 2
Capacitance range	<1 µF	100 pF to >100 µF
Temperature coefficient	Low	High
Capacitance changes with voltage?	No	Yes
Microphonic?	No	Yes

Class 2 parts are unsuitable for most applications beyond power supply bypassing because their capacitance changes significantly with temperature and applied voltage. They are also microphonic, which means that they convert sound or vibration into electricity.

Ceramic is extremely hard, but it is also brittle. Most failures of ceramic caps are due to mechanical shock to a circuit board from a drop or flexing of the board. Ceramic caps generally fail as shorts or a loss of capacitance.

If a chip capacitor is physically large enough, it should be marked with a code. Various codes are in use, but the Electronic Industries Alliance (EIA) code is the most popular for small-chip caps. Chips are marked as in Table 7-8.

TABLE 7-8 EIA Chip Cap Code

First Character	Second Character	Third Character
Company code	Letter code	Multiplier

The EIA code uses a letter to indicate value (Table 7-9). Uppercase and lowercase letters have different values, so pay attention to case. The multiplier is the exponent of 10 with units in picofarads, so a multiplier of 3 would be ×1,000 pF. A chip with code VQ3 would be a Vishay 3,900-pF cap.

TABLE 7-9 EIA Chip Cap Letter Code

Letter	Value	Letter	Value	Letter	Value	Letter	Value
A	1	M	3	Y	8.2		
B	1.1	N	3.3	Z	9.1		
C	1.2	P	3.6	a	2.5		
D	1.3	Q	3.9	b	3.5		
E	1.5	R	4.3	d	4		
F	1.6	S	4.7	e	4.5		
G	1.8	T	5.1	f	5		
H	2	U	5.6	m	6		
J	2.2	V	6.2	n	7		
K	2.4	W	6.8	t	8		
L	2.7	X	7.5	y	9		

Replacing Ceramic Capacitors

The biggest challenge in replacing ceramic chip capacitors is determining their capacitance value, voltage, and temperature coefficient. Some parts have no markings at all, which means you must find service data or take an educated guess. If the part is used for power supply bypassing, the value and temperature coefficient are not critical, and you may be able to measure another nearby bypass cap with the identical physical appearance as a clue.

If voltage isn't indicated on the part, you can measure the voltage with the power turned on as a clue. The voltage rating specifies the maximum allowed, so err on the side of choosing a replacement capacitor with a higher voltage rating.

Unlike the other types of caps, temperature coefficient (TC) is important with ceramic because there are so many extreme variations. Class 1 parts have low TCs, and Class 2 parts have high TCs. Class 1 parts should never be replaced with Class 2 parts, and your only indication is the TC. The lowest TC parts carry the code C0G (or NP0). If no code is present on the original, replacing with a C0G part is the best strategy for parts that are not bypass caps. If a different TC code is present on the original, order replacement parts with that code. The circuit may rely on a particular TC to compensate for the TC of another component.

Inductors

Inductors (Fig. 7-6), also called *chokes* and *coils*, are measured in henries. In practice, most parts are specified in microhenries (uH) or millihenries (mH). Because of their relative expense, and because they can pick up and radiate magnetic fields, inductors are avoided in design work whenever possible.

Inductors are easily distinguished from other components when their cores or coils are exposed. Encapsulated inductors look like resistors. They can fail open, shorted, or reduced value. The latter two happen when there's a short between turns of the windings. Use an ohmmeter to check for open inductors.

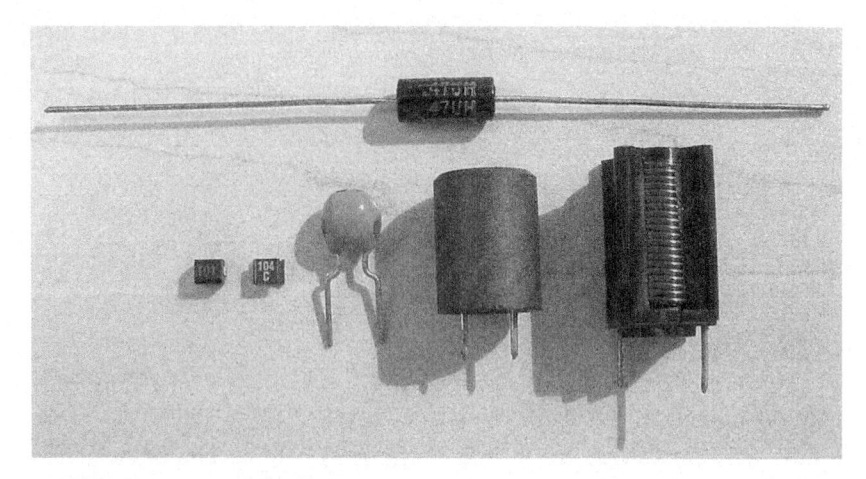

FIGURE 7-6 Variety of inductors.

Replacing Inductors

Besides inductance value, important parameters to consider during replacement are current rating, DC resistance, shielding, and self-resonant-frequency (SRF). SRF indicates the frequency where the parasitic capacitance of the part resonates with the inductance. Inductors no longer act like inductors above this frequency, so SRF can be regarded as an upper frequency limit for the part.

Often the only marking on an inductor is the value. On surface-mount parts, the marking is generally two value digits followed by a multiplier. The multiplier is the exponent of 10 with units in microhenries, so a multiplier digit of 3 would be ×1,000 µH. A chip with marking 223 would be 22 × 1,000 µH, or 22 mH.

When nothing other than the value is known, choose a shielded part with the same dimensions as the original. Although inductors aren't polarized components, some have a mark indicating the orientation of the internal windings. This can be important when multiple coils are in close proximity. Be sure to install the replacement with the orientation dot in the same position as the original.

Diodes

Diodes (Fig. 7-7) are recognizable by a band or bar on the package indicating the cathode terminal. They're common in both surface-mount and through-hole styles. Diodes can fail open or shorted. They can be checked using the diode test feature of a multimeter (see Chapter 6 for

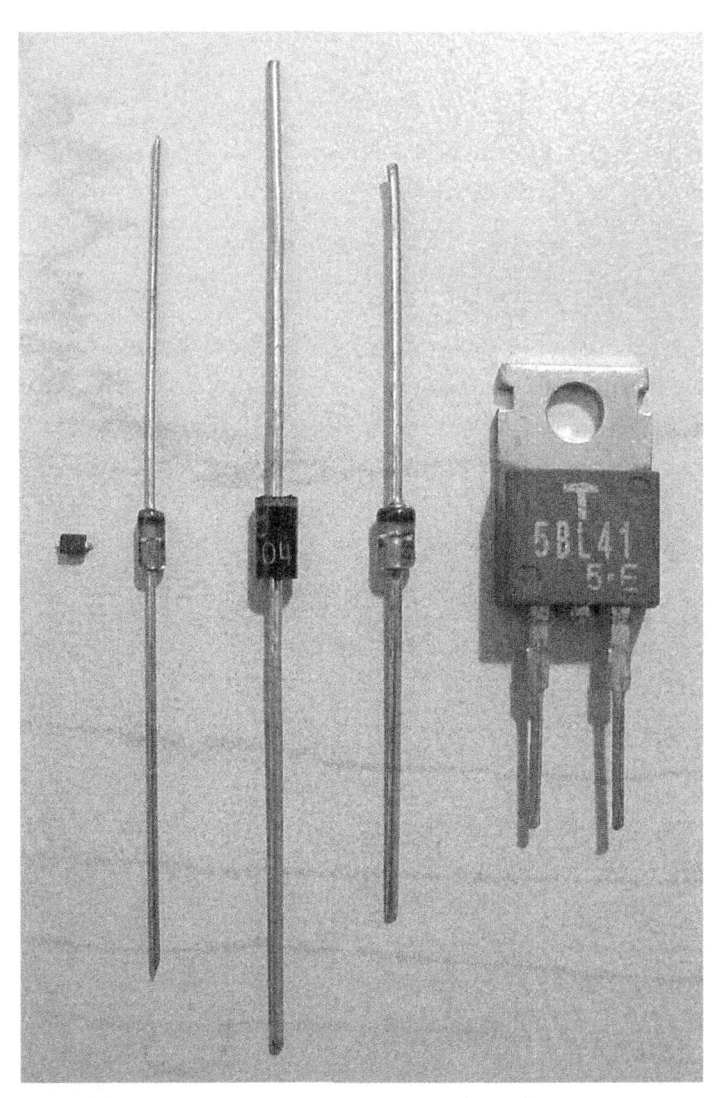

FIGURE 7-7 Variety of diodes: SMD, signal, rectifier, Zener, and TO-220-2 rectifier.

the procedure). Diode types include signal, rectifier, Schottky, Zener, and varactors.

For through-hole parts, the part number is usually marked on the body. For surface-mount parts, a code is usually used because of limited printing space. Refer to *The SMD Codebook* or other online resources to help identify parts marked with codes.

Signal Diodes

Signal diodes are general-purpose low-power diodes. If you can't identify the original, a good general-purpose signal diode is the 1N4148.

Rectifier Diodes

Rectifier diodes are used to convert AC to DC in power supply circuits. You'll usually find them in groups of two or four in a bridge configuration. Because rectifiers often handle high voltages and currents, important parameters for them are the current rating and peak reverse voltage.

If you can't identify the original, a good general-purpose rectifier diode is the 1N4004. It's a 1-amp diode with 400-volt peak reverse voltage. For higher power, the 1N5404 is a 3-amp diode with 400-volt peak reverse voltage.

Schottky Diodes

Schottky diodes have fast switching times and low forward voltage drops. This makes them ideal for switch-mode power supplies. Switch-mode power supplies are commonly used in many electronic products including computers, televisions, and AC adapters.

Commonly used Schottky rectifier diodes are the 1N58xx series, such as the 1N581x (1 amp) and 1N582x (3 amp) through-hole parts and the SS1x (1 amp) and SS3x (3 amp) surface-mount parts. Low peak reverse voltage is one of the limitations of Schottky devices. When substituting, pay close attention to the current rating and peak reverse voltage specs.

Zener Diodes

Zener diodes (or the related avalanche diodes) conduct electricity when they reach a specified voltage. This makes them useful for voltage references, power supply regulators, and overvoltage protection circuits. Zener voltages can be as low as 1 volt and as high as several hundred volts.

Zeners usually fail as a short, which is easily detected in or out of circuit with the diode test feature of a multimeter. When replacing Zener diodes, the two most important parameters are voltage and power dissipation. These two parameters must be matched with the original for a circuit to work properly. If you replace a part with one having higher power dissipation, the voltage will be low, and the voltage regulation will be degraded because the device will be operating at a lower bias current than it was designed for.

Varactor Diodes

Varactor diodes are used as electrically controlled capacitors in tuners and oscillators. They are just one of many specialty diode types that exist. They need an exact replacement for a circuit to work.

Transistors

Transistors (Fig. 7-8) are generally identifiable by their three-pin packages. Many voltage regulators and some diodes also use three-pin packages, so part numbers are the only way to tell them apart.

For through-hole parts, the part number is usually marked on the body. For surface-mount parts, a code is usually used because of limited printing space. Refer to *The SMD Codebook* or other online resources to help identify parts marked with codes.

FIGURE 7-8 Variety of transistors.

Transistor Packages

Transistors come in many different through-hole and surface-mount packages (Fig. 7-9). Sometimes the same transistor is offered with more than one packaging option. Larger packages provide better heat dissipation, either directly to the air or through a heat sink. The order of the pins can change with the package option, too.

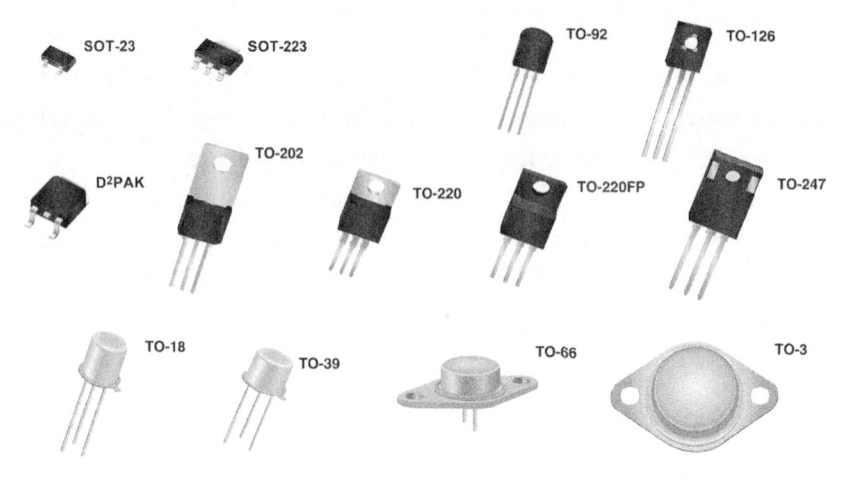

FIGURE 7-9 Transistor packages. *(Courtesy of Central Semiconductor Corp.)*

There are many types of transistors: NPN, PNP, and n-channel and p-channel MOSFETs (metal-oxide-semiconductor field-effect transistors), to name a few. The three pins of bipolar transistors (NPN or PNP) are called the *base, emitter,* and *collector*. The three pins of MOSFETs are called the *gate, source,* and *drain*. The most common types can be checked out of circuit using a multimeter. See the following sidebars for the procedures.

Transistor Failure Modes

Bipolar transistors usually fail with a shorted base to emitter or collector to emitter. MOSFETs usually fail with a shorted drain to gate or drain to source. If a MOSFET fails with a shorted drain to gate, this puts the drain voltage onto the gate, where it's likely to damage the drive circuitry, so the drive circuitry should be checked when replacing a MOSFET with shorted drain to gate. MOSFET failures tend to be dramatic, so usually if they look okay, they are.

TESTING BIPOLAR TRANSISTORS

NPN and PNP transistors can be checked out of circuit using the diode test feature of a multimeter. Table 7-10 shows the diode test results for good NPN and PNP transistors. Connect the meter probes as shown in the table, and compare your results against those in the appropriate transistor type column.

TABLE 7-10 Testing Bipolar Transistors

Positive Lead (Red)	Negative Lead (Black)	NPN Diode Test	PNP Diode Test
Base	Emitter	~0.6 volt	Over limit (OL)
Base	Collector	~0.6 volt	OL
Emitter	Base	OL	~0.6 volt
Collector	Base	OL	~0.6 volt
Emitter	Collector	OL	OL
Collector	Emitter	OL	OL

If you don't know whether a transistor is NPN or PNP or don't know which pins are which, you can still perform the test. Just try all six combinations of meter probe connections to pins of the device and see whether two of the results are about 0.6 volt and the rest are over limit (OL).

Note: This test verifies that the transistor is not shorted or open. It does not guarantee that the transistor is operating within all its design parameters.

TESTING ENHANCEMENT-MODE MOSFETS

n-Channel and p-channel enhancement-mode MOSFETs can be checked out of circuit using this procedure. Depletion-mode MOSFETS cannot be checked with this procedure but are rarely used in circuits.

Safety Note: MOSFETs are highly sensitive to ESD. ESD-safe practices should be used when handling them, such as using an antistatic mat and wrist strap. Use solder wick or an ESD-safe solder sucker when removing MOSFETs from circuit boards.

This procedure uses the diode test feature of a multimeter but requires that the open-circuit voltage applied by the meter is high enough to partially turn on the MOSFET. Meters that use a low open-circuit voltage (such as 1.5 volts) will not work. Open-circuit voltages of 3–4 volts are good.

You'll need to know which pins are the gate, source, and drain to do the test, so it's helpful to first identify the part. Most MOSFETs in a TO-220 package have the leads in gate, drain, source (GDS) order left to right.

For n-Channel Enhancement-Mode MOSFETs

1. Connect the meter's negative lead (black) to the MOSFET's source pin.
2. Use a clip lead to momentarily short the MOSFET's gate pin to source pin. This turns the transistor off. The MOSFET's gate capacitance is sufficient to hold the voltage afterward.

3. Connect the meter's positive lead (red) to the MOSFET's drain pin. The meter should show OL, indicating the transistor is off.
4. Remove the meter's positive lead (red) from the drain pin and momentarily touch it to the gate pin. This turns the transistor on. The MOSFET's gate capacitance is sufficient to hold the voltage afterward.
5. Connect the meter's positive lead (red) back to the MOSFET's drain pin. The meter should show very close to 0 volts, indicating that the transistor is on.

For p-Channel Enhancement-Mode MOSFETs
1. Connect the meter's positive lead (red) to the MOSFET's source pin.
2. Use a clip lead to momentarily short the MOSFET's gate pin to source pin. This turns the transistor off. The MOSFET's gate capacitance is sufficient to hold the voltage afterward.
3. Connect the meter's negative lead (black) to the MOSFET's drain pin. The meter should show OL, indicating that the transistor is off.
4. Remove the meter's negative lead (black) from the drain pin and momentarily touch it to the gate pin. This turns the transistor on. The MOSFET's gate capacitance is sufficient to hold the voltage afterward.
5. Connect the meter's negative lead (black) back to the MOSFET's drain pin. The meter should show very close to 0 volts, indicating that the transistor is on.

Transistor Substitution

Numerous transistor substitution tools are available online to help you when the original part is difficult or impossible to find. Some of these are alltransistors.com and nteinc.com. In many cases, a general-purpose transistor will suffice for a repair. Table 7-11 shows some top picks. Keeping some of these on hand can be a big time saver.

TABLE 7-11 General Purpose Transistors for Substitution

Transistor	Description
2N3904	NPN, TO-92 package EBC, 200 mA, 40 volts
2N3906	PNP, TO-92 package EBC, 200 mA, 40 volts
TIP31	NPN, TO-220 package BCE, 3 amps, 60 volts
TIP32	PNP, TO-220 package BCE, 3 amps, 60 volts
2N7000	n-Channel enhancement-mode MOSFET, TO-92 package SGD, 200 mA, 60 volts

Voltage Regulators

Three-terminal linear voltage regulators (Fig. 7-10) look like transistors but are integrated circuits. They deserve special coverage because of how often they're used. The most common package is the through-hole TO-220, but voltage regulators come in a range of packages from surface mount to the big TO-3 can depending on power dissipation requirements.

FIGURE 7-10 Variety of voltage regulators.

Three-terminal regulators provide a clean and stable voltage source from a higher voltage source. Regulators come in both positive and negative versions and in fixed and adjustable types. The LM78xx series of positive and LM79xx series of negative regulators are extremely popular. The "xx" represents the regulated output voltage; for example, LM7905 is a −5-volt regulator. The LM317 (positive) and LM337 (negative) are a popular series of adjustable regulators. The voltage of adjustable regulators is set using two external resistors.

Checking Regulators

Three-terminal regulators can be checked in normal operation by measuring the DC voltages of the three terminals. For fixed regulators, the three terminals are input, reference, and output. Usually, the reference pin is connected to ground. For positive regulators, you should see a voltage at the output pin that is the specified regulated voltage higher than the reference pin. The voltage at the input pin should be at least 2.5 volts higher than the output pin but not more than 35 volts higher than the reference pin. For example, for an LM7812, you might measure 20, 0, and 12 volts for the input, reference and output pins, respectively. For negative regulators, you should see the negative voltage equivalent of these results.

For adjustable regulators, the three terminals are input, adjust, and output. For these, you should see a 1.25-volt difference between the adjust and output pins. You should also see at least a 2.5-volt difference between the input and output voltages.

If the measured voltages don't match what you expect, the regulator may or may not be at fault. There may be a problem in the circuit providing the input to the regulator or a short in the circuit using the regulated output voltage. A typical failure mode for three-terminal regulators is an input-to-output short circuit, but open outputs are also possible.

Integrated Circuits

Integrated circuits (Fig. 7-11), also called ICs, are easy to spot by their black plastic packages having many pins. For large parts, the part number is usually marked on the body. For small surface-mount parts, a code is usually used because of limited printing space. Refer to *The SMD Codebook* or other online resources to help identify parts marked with codes.

FIGURE 7-11 Variety of integrated circuits.

IC Packages

ICs come in many different through-hole and surface-mount packages. Figure 7-12 shows some of the more common types. Sometimes the same IC is offered with more than one packaging option, which is useful to know when ordering a replacement.

Pin 1 is usually marked with a dot, and the pin numbers increase as you travel counterclockwise around the perimeter of the IC. On dual in-line packages (DIPs), there's an indentation on the pin 1 side of the chip (Fig. 7-13).

IC Testing

Because ICs are often complex and there are so many different kinds, there's no generic procedure for testing them. See Chapter 8 for troubleshooting IC circuits.

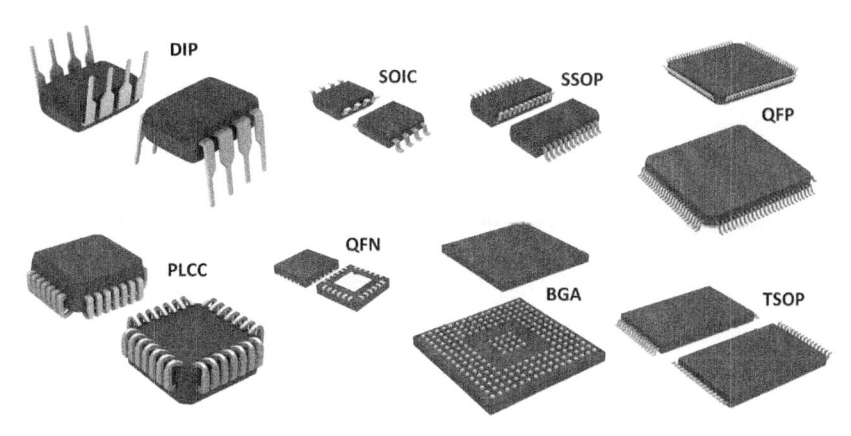

FIGURE 7-12 IC packages.

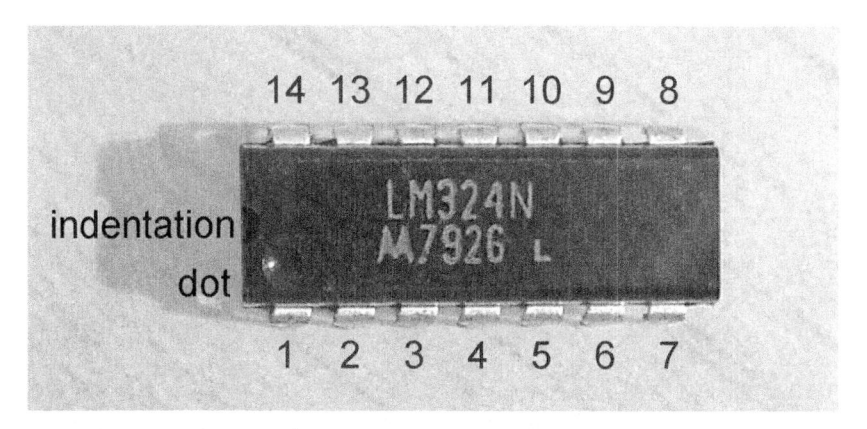

FIGURE 7-13 IC pin numbering.

IC Substitution

IC substitution is possible in some cases when the original part is difficult or impossible to find. In addition to transistors, NTE Electronics provides substitutes for some ICs such as operational amplifiers (op-amps), digital logic, audio power amps, microprocessors, display drivers, and motor controllers. Often the manufacturer's datasheet will include parts that may be acceptable substitutes but have different temperature ranges, speed grades, or maximum power supply ratings.

Varistors

Varistors (Fig. 7-14) are voltage-dependent resistors used for circuit protection. Most varistors are metal-oxide varistors (MOVs). They're commonly found in power supply circuits or surge protectors. For normal operating voltages, they have a high resistance value and do not affect circuit operation. With high voltage spikes, from lightning, for example, they have a low resistance value that shorts the spikes to ground.

Varistors lead an inherently precarious existence and are damaged when the energy of the spikes exceeds their ratings. They often fail dramatically. It's likely that the energy surge that caused a varistor to fail also damaged other components, so check nearby fuses and rectifier diodes when replacing varistors.

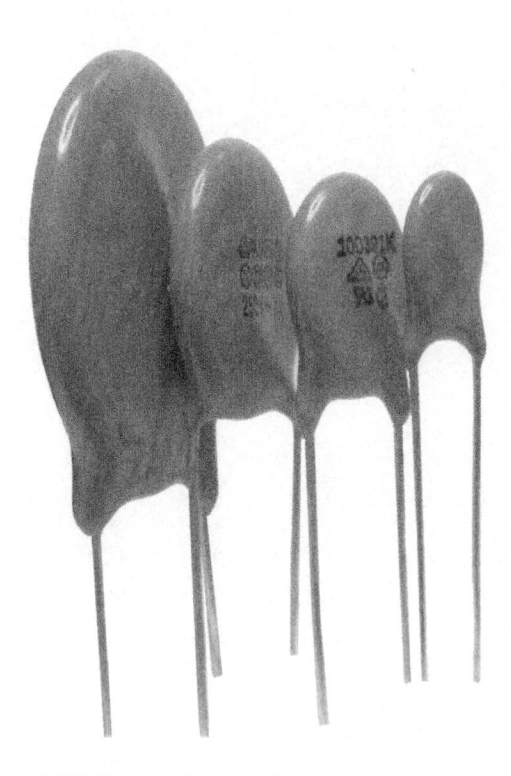

FIGURE 7-14 Variety of varistors.

Varistor Identification and Replacement

Identifying a varistor requires detective work because every manufacturer uses its own marking system. Google the part numbers found on the part. Some of the characters are likely to be date codes and won't help the search, so omit them if you think they're date codes. Knowing the manufacturer can help with the search. Every manufacturer has a distinct logo that's likely to be stamped on the part. Databases of electronic component manufacturer logos are readily available online.

When replacing a varistor, important parameters include the nominal voltage rating, peak current, energy rating, and capacitance. If substituting, make sure that the nominal voltage rating makes sense for the application. A 35-volt varistor would not be used across the power lines in a 120-VAC surge protector.

Thermistors

Thermistors (Fig. 7-15) are temperature-dependent resistors used for circuit protection and temperature compensation. Negative temperature coefficient (NTC) versions have decreasing resistance with increasing temperature. Positive temperature coefficient (PTC) versions have increasing resistance with increasing temperature.

FIGURE 7-15 Variety of thermistors.

Some PTC thermistors are used as resettable fuses. As the circuit gets hot, the thermistor reduces the voltage or current to it, protecting it. Thermistors usually fail as open circuits but can fail as a parameter change or short circuit.

Thermistor Identification and Replacement

The two most common markings on a thermistor are NTC or PTC and a number representing the nominal resistance at room temperature (25°C). The resistance could be shown directly or in digits plus multiplier format.

Identifying a thermistor requires detective work because every manufacturer uses its own marking system. Google the part numbers found on the part. Some of the characters are likely to be date codes and won't help the search, so omit them if you think they're date codes. Knowing the manufacturer can help with the search. Every manufacturer has a distinct logo that's likely to be stamped on the part. Databases of electronic component manufacturer logos are readily available online.

Once you know the nominal resistance, an ohmmeter reading will tell you whether the part is likely to be good or not. When replacing a thermistor, important parameters include whether it's NTC or PTC and the nominal resistance. For NTC thermistors, the B (or beta) value is an important specification. For PTC thermistors used as resettable fuses or heaters, the transition temperature is most important.

Fuses

Fuses (Fig. 7-16) are normally easy to identify and test. If visual inspection doesn't show that a fuse is blown, check it with an ohmmeter or continuity test. The current rating of a fuse and the fuse series identifier are usually stamped on the fuse. The series, such as AGC or AGX, defines the physical size, speed, and voltage rating.

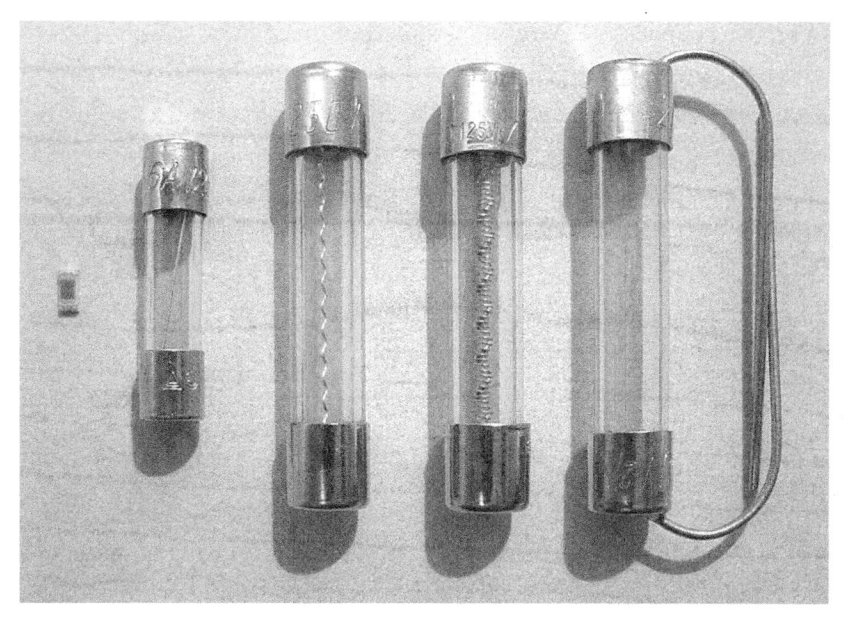

FIGURE 7-16 Variety of fuses: surface mount, 5- × 20-mm cartridge, ¼- × 1¼-inch cartridge, ¼- × 1¼-inch slow-blow cartridge, pigtail.

Fuse Failure

Fuse filaments can break from mechanical shock such as from dropped equipment. Power surges can blow fuses with or without damage to other components. Most of the time, a blown fuse means a component failure in or affecting the power supply circuit. Check the power supply diodes for opens or shorts. Other suspects are shorted power transistors and tantalum bypass capacitors anywhere in the device.

Fuse Replacement

When replacing a fuse, the important electrical parameters are current rating, speed, and voltage rating. Never replace a fuse with one with a current rating higher than the original or you risk equipment damage or creating a safety hazard. Fuses can have speeds of fast or slow-blow. Slow-blow fuses are more immune to short-duration current surges. Their filament designs are more complex, such as being wound around an insulating core or having multipart internal elements. The voltage rating of a fuse comes into play after a fuse blows, to ensure that no electrical arcing across the fuse gap occurs. It's okay to use a fuse with a higher voltage rating than the original but not lower.

The most common cartridge fuse sizes for electronics are the ¼ × 1¼ inch (diameter × length) and the 5 × 20 mm. If you find yourself looking to replace a missing fuse where the current rating can't be found but believe the equipment to be otherwise working, put in a test fuse and measure the current using a multimeter. Make your measurement with the equipment at maximum load. Choose a fuse current rating that's 1.33 times the measured current to provide a margin against nuisance blows.

Pigtail fuses look like cartridge fuses but with a wire soldered on each end. You can solder wires on the ends of a cartridge fuse as a substitute, but you'll need to solder fast or you'll melt the filament. Using a large soldering iron tip and high heat will improve your chances of success.

Switches

Because switches (Fig. 7-17) are mechanical devices, they're a common source of problems. Switches can fail because of wear, oxidation, or electrical arcing of contacts. Check switches with an ohmmeter or continuity test.

Switches come in so many varieties that an entire chapter could be devoted to them. Mouser Electronics has more than 150,000 to choose from; Digi-Key Electronics has even more. In spite of that, you may need to resort to substitution because an exact replacement isn't available.

Table 7-12 gives examples of parameters to consider when replacing a switch. Most are easy to understand, but contact form (pole and throw terminology) and switch function are explained in more detail later.

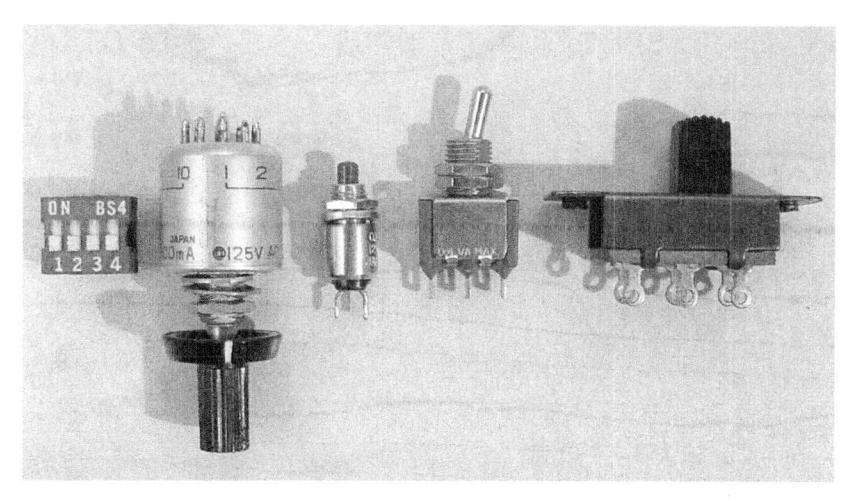

Part Identification, Testing, and Substitution | 133

FIGURE 7-17 Variety of switches: DIP, rotary, pushbutton, SPDT, DPDT.

TABLE 7-12 Switch Parameters

Parameter	Examples
Type	Toggle, pushbutton, rocker, rotary, slide
Mounting style	Panel mount, through-hole, surface mount
Contact form	SPST, SPDT, DPDT
Switch function	ON-ON, ON-OFF-MOM, (ON)-NONE-(ON)
Termination style	PC pin, solder lug, screw
Illumination	None, LED, incandescent, neon
Current rating	1 amp, 10 amp
Voltage rating	120 VAC, 12 VDC

Pole and Throw Terminology

A switch must have at least two terminals, but often switches have three, six, or even more terminals. The connections between all those terminals are described using poles and throws (Fig. 7-18). Poles are the number of subswitches the switch contains. Throws are the number of wired positions each subswitch provides.

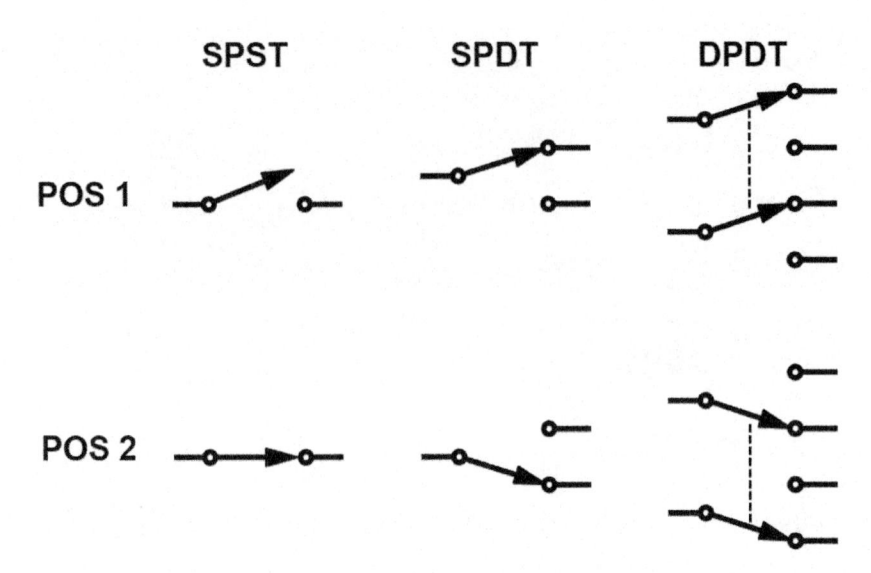

FIGURE 7-18 Pole and throw terminology: SPST, SPDT, DPDT.

A single-pole, single-throw (SPST) switch is as simple as it gets. It's got a single subswitch with a single wired position. SPST switches have two terminals that are either connected (on) or not (off) depending on which of the two switch positions is selected.

Next in the lineup is a single-pole, double-throw (SPDT) switch. It has a single subswitch with two wired positions. SPDT switches have three terminals, and the center terminal is either connected to one or the other outer terminal depending on which of the two switch positions is selected. It can be wired as an on-off switch by not using one of the outer terminals.

Double-pole, double-throw (DPDT) switches have two subswitches, each with two wired positions. These have six terminals arranged in a 2 × 3 grid, and the center terminal of each subswitch is either connected to one or the other outer terminal depending on which of the two switch positions is selected.

Switches with four poles and three throws (4P3T) and even higher exist. When replacing any switch beyond SPST, it's important to wire the replacement the same way as the original. Take a picture with your phone beforehand as an insurance policy.

Switch Function

Most toggle switches have two positions, but some have a third position in the center. Switch function is described by two or three terms separated by hyphens, such as ON-ON or ON-OFF-MOM, where MOM stands for "momentary" and indicates a spring-loaded position. Different manufacturers use slightly different terminology to describe the functions of their switches. A term in parentheses also means momentary, so ON-OFF-(ON) is the same as ON-OFF-MOM.

Try Contact Cleaner

Contact cleaner such as CAIG DeoxIT D5 can often revive switches with contact oxidation problems. The trick is directing the spray into the contacts without getting it everywhere else. After spraying, work the switch control back and forth 10 or more times to remove the layer of oxidation from the contacts.

Replacement and Substitution

Using an exact replacement solves the problem of trying to match a myriad of switch specs. Most switches have the manufacturer, part number, and current and voltage ratings on them. If you're not able to find the original, your next-best chance is a substitute from the same manufacturer. The major electronics suppliers usually put spec sheets on their websites that cover an entire series of related switches. Often the only difference between old and new parts is RoHS compliance or some parameter that doesn't matter to you.

WHAT IS ROHS COMPLIANCE?

RoHS stands for Restriction of Hazardous Substances. RoHS, also known as Directive 2002/95/EC, originated in the European Union and restricts the use of specific hazardous materials found in electronics. Banned substances include lead, mercury, and cadmium. The restricted materials are hazardous to the environment and pollute landfills.

When substituting, make sure that the replacement has the same or higher voltage and current ratings. Parts often carry separate (and radically different) AC and DC ratings. In addition, inductive load ratings may be given. Observe the rating that applies to whether the switch will be used for AC or DC and whether or not you have an inductive load such as a motor. Microswitches may specify their rating in volt-amperes. In such cases, you multiply the application voltage by the application current to convert one rating system to the other.

DOUBLE-POLE POWER SWITCH

A simple way to increase the reliability of a power switch is to use a double-pole version of the switch and wire the two sections in parallel (Fig. 7-19). For example, replace an SPST switch with a DPST switch.

FIGURE 7-19 Double-pole power switch.

Transformers

Most transformers are used in power supply circuits. These power transformers (Fig. 7-20) are used to convert AC voltage to a lower (step-down) or higher (step-up) level as well as provide electrical isolation for safety. Signal transformers are sometimes used to provide isolation or impedance transformation in radio-frequency (RF) circuits, telephone equipment, or vintage audio gear.

FIGURE 7-20 Variety of power transformers.

Transformers are constructed of wire windings on a shared magnetic core. The two input wires correspond to the primary winding; the two output wires correspond to the secondary winding. Some transformers have multiple secondary windings that are used to supply AC power to separate power supply circuits. A center-tapped secondary has a third wire connected to the electrical center of the secondary winding. It's used to reduce the number of diodes needed in the rectifier circuit or to produce positive and negative DC supplies.

Transformer Failure

Power transformers usually fail open-circuit, but shorts between turns of the windings are also possible, resulting in higher or lower output voltages depending on whether the shorted turns are within the primary or secondary. Shorts between primary and secondary windings are extremely rare because of the construction of power transformers, which are designed to prevent this potentially fatal failure mode. Transformers often have a built-in thermal fuse to prevent a fire in the event of overheating. When these fail, the winding will appear open, and you'll need to replace the transformer.

The simplest test for a transformer is to check for open circuits with an ohmmeter. If none of the windings are open, the next test is to measure the AC voltage at the secondary output. Use caution when working around potentially lethal voltages that may be present at transformer windings. Use of an isolation transformer is recommended. The secondary voltage should be close to its rated output voltage, usually stamped on the transformer.

Power surges can blow transformers, but a blown transformer can mean a component failure in or affecting the power supply circuit. Check the power supply diodes for opens or shorts. Other suspects are shorted power transistors and tantalum bypass capacitors anywhere in the device.

Transformer Replacement and Substitution

When replacing power transformers, important parameters include the primary voltage rating, secondary voltage rating(s), and the secondary current rating(s). The primary and secondary voltage ratings should closely match the original. The secondary current ratings should be equal to or higher than the original.

The output voltage of transformers changes with loading. The rated voltage is what you can expect at the rated current. Under no-load conditions, the output voltage can be considerably higher. If you substitute a same-voltage but higher-current transformer, the output voltage will be higher than the original. There's no simple way to know the point at which this might cause a problem, so it's best to avoid substitutions with more than a two-to-one current rating increase.

Relays

Relays (Fig. 7-21) are electrically operated switches. Because of this, they share many of the same parameters as switches, such as contact current and voltage ratings and contact form (SPDT, DPDT). Relays have two terminals for an electromagnetic coil; the others are switch terminals. When the coil is energized with electricity, it activates the mechanical switch inside. Many relays have multiple switches inside, controlled by a single electromagnetic coil. Relays allow a low control voltage to control a high-powered switch circuit.

Because relays are mechanical devices, they're a common source of problems. Relays can fail because of wear of the mechanism or electrical arcing of contacts.

FIGURE 7-21 Variety of relays.

Relay Testing

To check a relay out of circuit, you'll need a way to energize the coil to activate the internal switch. Most relay coils operate on a DC voltage such as 5 or 12 volts, which is usually printed on the relay. A DC power supply comes in handy for providing this. The polarity of the applied DC doesn't matter, except where the relay has an internal diode. Normally, external diodes are used, but some relays have them built in. Diodes are used across relay coils to short out the large negative inductive voltage spikes that occur when turning off the coil voltage.

Once you have a way to turn the coil voltage off and on, use an ohmmeter or continuity tester to check the operation of the relay switch. If you don't hear the relay click as you turn it off and on, the relay is stuck in one position and should be replaced. Relay contact pins are usually labeled "NO" (normally open), "C" (common), or "NC" (normally closed), where *normal* means no power is applied to the coil. There should be a connection between the C and NC terminals when no power is applied and between the C and NO terminals when power is applied.

It's not always necessary to fully desolder a relay to test it. If you can desolder the two coil pins so that they no longer make electrical contact with the PC board traces, you can test the relay in place. You can carefully solder bits of wire to the desoldered coil pins to make connection to the power supply leads easier.

Replacement and Substitution

Table 7-13 gives examples of parameters to consider when replacing a relay.

The contact form of relays is similar to that of switches. The difference is that for relays, it's necessary to define the switch position when there's no power applied to the coil. An SPST-NO relay has two contact pins that are open when no power is applied and closed when power is applied. Relay manufacturers use the form system to describe this, where the number refers to the number of switch poles and the letter refers to the unpowered state.

FIGURE 7-13 Relay Parameters

Parameter	Examples
Contact form	SPST-NO (1 Form A), SPST-NC (1 Form B), SPDT (1 Form C)
Contact current rating	1 amp, 10 amps
Contact voltage rating	120 VAC, 12 VDC
Coil voltage	5 VDC, 12 VDC
Coil resistance	100 ohms, 1,000 ohms

The contact current and contact voltage ratings pertain to the switches inside the relay. They should be the same or greater than the original. The coil voltage should precisely match the original for proper switching. Ideally, coil resistance should also match the original, but this isn't always possible or necessary. Having a coil resistance that is higher than the original will work in almost all applications. In those rare cases where a higher resistance causes a problem, you can solder in a parallel resistor to bring the coil resistance down.

Using an exact replacement solves the problem of trying to match multiple relay parameters, especially the exact pin configuration. Most relays have the manufacturer, part number, coil voltage, and contact current and voltage printed on them. If you're not able to find the original, your next-best chance is a substitute from the same manufacturer. The major electronics suppliers usually put spec sheets on their websites that cover an entire series of related relays. Often the only difference between old and new parts is RoHS compliance or a parameter that may not matter to you, such as the relay height.

Analog Meters

Analog meters (Fig. 7-22) are increasingly rare, but if you're repairing a piece of beloved vintage audio equipment or a power supply, here's how to test, repair, and replace the faulty parts. Use an ohmmeter to test the resistance of the meter. You'll need to disconnect at least one of the meter terminals from the circuitry for this. If you measure an open circuit, the meter will need to be replaced.

FIGURE 7-22 Analog meter.

Stuck Needle Fixes

If the meter is stuck at a particular position, remove the clear plastic front cover of the meter to expose the meter movement. Using a toothpick, very gently try to move the needle over its range. The needle should return to the zero position by its internal spring. If the needle rubs against the scale, it may have gotten bent from an equipment drop. Gently bend it away from the scale to see if that fixes the problem.

If the needle is not rubbing against the scale but just seems to be binding along some part of its range, the next thing to try is readjusting the pivot screw. If there's sealant on the screw, use acetone on a cotton swab to remove or soften it. Using a jeweler's screwdriver, very gently loosen the pivot screw a fraction of a turn. See if this fixes the problem. Do not loosen the screw too much or the movement will come apart. Use nail polish to reseal the screw after the fix.

In some cases, the problem is that the needle is out of proper alignment because of a drop. You'll need to loosen the pivot screw a few turns, properly position the needle in its movement, and then tighten the screw. Move the needle over its range after each adjustment. If the needle starts to bind, you'll need to loosen it a bit.

Another method for fixing stuck meter needles is the soldering iron method. This may work if the problem is gummed lubricant on the needle pivot. A hot soldering iron tip is held on the pivot screw long enough to burn away the gummed lubricant.

Meter Replacement and Substitution

Ideally, an exact replacement can be purchased from the equipment manufacturer or an eBay seller. If not, you'll need to consider substitution. When substituting, the most important thing to know is that there's usually no relationship between what's on the meter scale and the actual current or voltage driving the meter. A power supply meter may show a scale of 0 to 50 VDC, but the meter itself could be a 200-µA full-scale movement. This means that for replacement, you would need to find a 200-µA meter that also happens to have a scale showing 50 volts. Finding such a meter that's also the right size and price is like winning the lottery. The alternative is to buy a substitute with the right full-scale current and relabel the meter scale.

Most meters respond to DC current, but some respond to DC voltage. To determine what you need, connect your multimeter in place of the meter, and measure the DC voltage with the equipment operating to create a full-scale condition. Now put a 1,000-ohm resistor in parallel with the meter. If the reading doesn't change, you need a voltmeter with the indicated full-scale voltage. If the reading does change, this means that you need an ammeter, not a voltmeter. Put your multimeter in the DC current mode to measure the full-scale value. After replacing a meter, you may need to calibrate the meter level adjustments of your equipment for accuracy.

AC Adapters

AC adapters (Fig. 7-23), also known as *wall warts*, *wall chargers*, or *power bricks*, have become ubiquitous in electronics. This section will tell you how to test them, repair them, and find a substitute if necessary.

FIGURE 7-23 AC adapters.

AC Adapter Basics

If a product appears powerless, your first suspect (after making sure that it's plugged in and turned on) should be the AC adapter. Most AC adapters provide a DC output, but there are a few that provide an AC output.

There are two types of DC output adapters: unregulated and regulated. Unregulated adapters are just a transformer with diode rectifier and maybe an electrolytic capacitor. The output voltage of these changes dramatically depending on how much current is being drawn. A 12-volt unregulated adapter might measure 12 volts at the rated current and 15 volts or more with no load. Regulated adapters can be linear or switch mode. Almost all regulated adapters nowadays are switch mode. These are identifiable by their comparatively small size, light weight, and wide input voltage range (100–240 VAC). Regulated adapters provide an output voltage that's almost the same regardless of how much current is being drawn. To meet energy compliance requirements, some regulated adapters go into a low-power mode when the current drops below a certain value, and the no-load output voltage can be low or irregular.

Testing AC Adapters for a Bad Cable

The most likely problem with an AC adapter is a bad connection in the cable. The high-stress points are usually at the two ends, where the cable exits the body of the adapter and at the output connector. With the adapter plugged into wall power, use a multimeter to measure the voltage at the output connector (AC volts for adapters with AC outputs, DC volts for adapters with DC outputs). If you measure 0 volts, try wiggling the cable in every direction and pushing and pulling it at each stress point to produce a moment of power from your adapter. If you succeed, you've located the bad connection.

If you can't produce a moment of power, unplug the adapter from the wall, and measure the resistance at the output connector. If it's an open circuit, the evidence points to a bad connection in the cable.

Testing AC Adapters for Correct Output Voltage

If you were able to measure more than 0 volts at the output connector, the next step is to check for the correct output voltage. The nominal output voltage and current rating are almost always printed on the adapter.

If the adapter is the AC output type, set your meter to AC volts. You should measure something higher than the nominal voltage because you're measuring the no-load voltage. The output voltage will decrease as more current is drawn because of the internal impedance of the transformer windings and other factors.

If the adapter is the DC unregulated type (it's big and heavy), set the meter to DC volts. Once again, you should measure something higher than the nominal voltage because you're measuring the no-load voltage. If the adapter is the DC regulated type, the test is more involved.

Measure the DC voltage. If it's very close to the nominal voltage, that's a good sign. If it's not, the adapter could be in a low-power mode and might still be okay. Next, you want to measure it at or near its rated output current. Here's where a dummy load or assortment of power resistors comes in handy. To test at rated load, you need a resistor with a value calculated using Ohm's law (see Chapter 3). For example, if you're testing a 12-volt, 0.5-amp adapter,

$$R = V/I = 12/0.5 = 24 \text{ ohms}$$

Don't use a resistor with a lower resistance or you could damage a working adapter. For simple testing, a somewhat higher value resistor will usually reveal whether the adapter is working or not. Calculate the power that the resistor will need to handle (also in Chapter 3):

$$P = V^2/R = 12^2/24 = 6 \text{ watts}$$

Use a resistor with this power rating or higher. If the measured voltage is very close to the nominal voltage for this test, the adapter is assumed to be good.

AC Adapter Repair

Determine whether the AC adapter case is easily opened or not. See the section "Taking It Apart" in Chapter 8 for more information on this.

If the case is not easily opened and a bad cable connection is suspected, cut off the output connector about an inch from the connector. Strip back the insulation on the end of the cable and retest the adapter for voltage. If you have voltage, buy and install a new power connector on the end of the cable. See the sidebar "Replacing DC Power Connectors."

If the case is easily opened or you don't have voltage on the stripped cable end from the preceding step, you'll want to open the case. Once opened, use your multimeter to perform a continuity test on the two cable wires from inside the case to the stripped cable end or output connector. If the cable continuity test is good, look for bulging, burst, or leaking electrolytic capacitors. If you have an ESR or capacitance meter, check all the electrolytics. If the caps all look good, refer to Chapter 8 for more advanced troubleshooting guidance.

REPLACING DC POWER CONNECTORS

Solder-on DC power connectors are available in dozens of sizes to replace the molded-on originals. Unfortunately, there are so many possible sizes that you'll need to carefully measure your old connector to get the right match. Having digital calipers is very helpful for this. You'll need to measure the outer diameter, inner diameter, and length. If you have a set of machinist drill bits, you can use them to measure the inner diameter by finding the bit that just fits. Use the back end of the bit, not the front. The length of the replacement barrel should be at least as long as the original.

The photos in Figure 7-24 show the steps for soldering on a DC power connector. Be sure to get the polarity correct. Usually, the center of the connector is positive, but not always. The AC adapter case and/or the device case should show this. Use your multimeter to check the polarity of the wires before you solder them to the connector and again after you're done.

148 | The Electronics Repair Cookbook

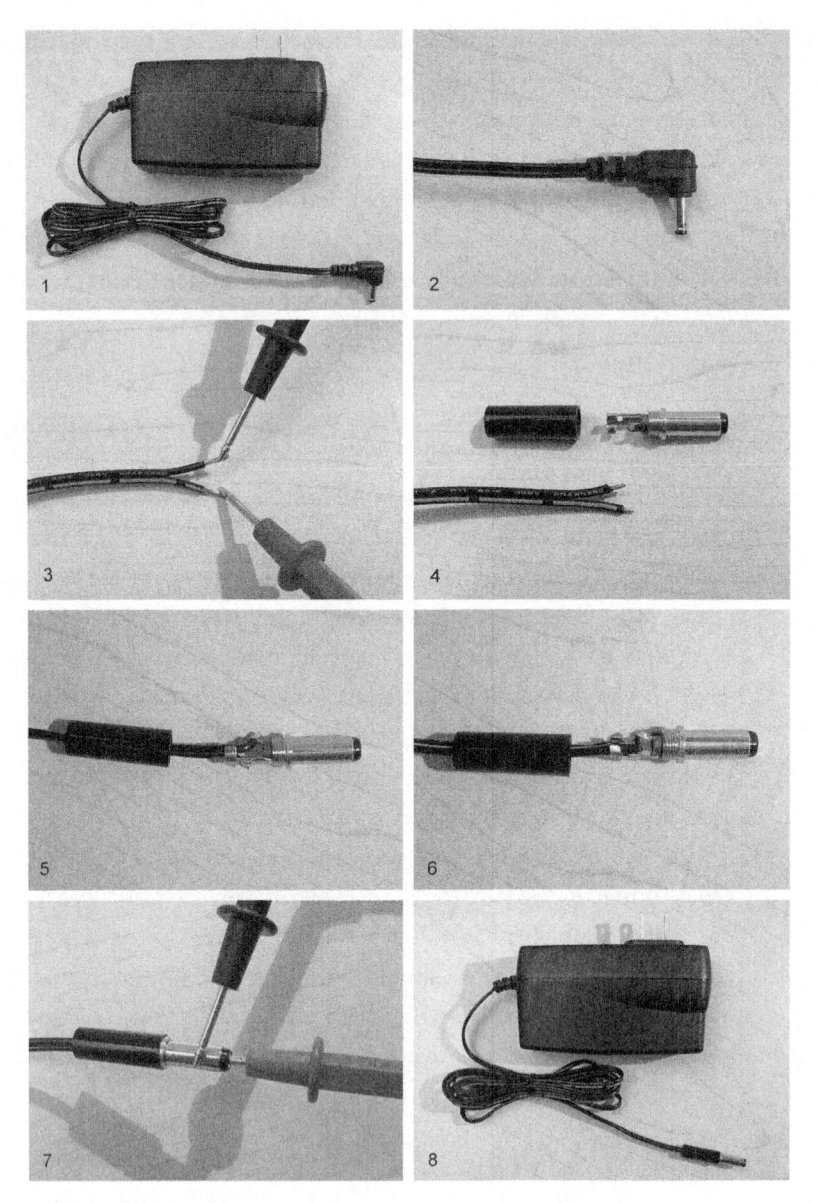

FIGURE 7-24 Soldering on a DC power connector and testing.

AC Adapter Substitution

When searching for a substitute AC adapter, you'll need to match the voltage type (AC or DC), voltage, current, and connector. If the original adapter was regulated, the substitute should be too. Unregulated DC adapters may be replaced with regulated or unregulated DC adapters.

The current rating of the substitute adapter depends on the type of adapter. Table 7-14 shows the acceptable range. With unregulated adapters, the voltage changes with load, and using an adapter with a much larger current rating runs the risk of providing an output voltage that's too high. With regulated adapters, the voltage remains the same with load, and there's no overvoltage issue with higher current specifications. The only danger in using an adapter with a lot more current capacity than you need is if there's a short circuit in the device. The extra current can cause much more damage.

TABLE 7-14 AC Adapter Substitution Current Rating

Original Adapter	Substitute Adapter	Current Rating (Original = 100%)
AC	AC	100–150%*
DC unregulated	DC unregulated	100–150%*
DC unregulated	DC regulated	100% or higher
DC regulated	DC unregulated	n/a
DC regulated	DC regulated	100% or higher

Assumes 25 percent transformer regulation producing 8 percent voltage increase.

Sometimes the biggest challenge in finding a substitute adapter is finding one with the right power connector dimensions and polarity. Frequently, the full connector specs aren't provided by the seller. If you find an AC adapter that meets your electrical requirements, you can always solder on a replacement DC power connector that meets your connector requirements. See the sidebar "Replacing DC Power Connectors."

Component Test Summary

Table 7-15 summarizes what equipment to test each component type with. Most components can be checked with a standard multimeter. See the preceding sections for details.

TABLE 7-15 Component Test Summary

Component	Test with
Resistors	Ohmmeter (multimeter)
Potentiometers	Ohmmeter (multimeter)
Capacitors	ESR meter or capacitance meter
Inductors	Ohmmeter (multimeter) or inductance meter
Diodes	Diode test (multimeter)
Transistors	Diode test (multimeter)
Voltage regulators	DC voltmeter (multimeter)
Integrated circuits	Troubleshooting
Varistors	Ohmmeter (multimeter)
Thermistors	Ohmmeter (multimeter)
Fuses	Ohmmeter or continuity test (multimeter)
Switches	Ohmmeter or continuity test (multimeter)
Transformers	Ohmmeter (multimeter), AC voltmeter (multimeter)
Relays	Power supply and ohmmeter (multimeter)
Analog meters	Ohmmeter (multimeter)
AC adapters	DC voltmeter (multimeter), ohmmeter (multimeter)

CHAPTER 8

General Repairs and Troubleshooting

This chapter provides repair guidance that can be applied to any electronic device. Chapter 9 provides additional information for frequently repaired devices.

Is It Worth Fixing?

There's no simple answer to whether or not something is worth fixing. It usually makes sense to at least check for an easy repair. Take off the cover and see what's inside. The problem may be obvious, and the repair may take very little effort.

Even if a device isn't worth fixing, it may have parts value. There's probably somebody looking for a part that's still good on your device. It could be a remote control, stand, set of knobs or feet, AC adapter, or electronic subassembly.

Poor Candidates for Repair

Some repairs are notoriously bad investments of your time. These include

- Lightning-damaged equipment
- Liquid-damaged/submerged equipment

All the damage done by lightning may not show up right away. The high voltage induced by nearby lightning strikes into AC power and cable lines (TV, internet, phone) can weaken or degrade semiconductors so that they're poised for future failure. You might think you've fixed a lightning-damaged device, but you'll be working on it again soon.

It's surprising how much damage liquids can cause to multilayer PC boards with surface-mounted microelectronics, even ordinary drinking water. Soda and saltwater are worse. If the power was on when the spill occurred, the damage is multiplied. If a simple cleanup of a spill doesn't fix the problem, a quick repair is unlikely.

How Did It Break?

Knowing how a device broke provides valuable clues about what's wrong.

- As mentioned earlier, devices damaged during a thunderstorm or from liquids are poor candidates for repair.
- Devices with problems that come and go are especially challenging. See the section "Intermittent Problems" later in this chapter.
- Did the problem gradually get worse? Failing electrolytic capacitors are the leading cause of gradually worsening symptoms. Test them with an ESR meter or the parallel-cap technique (explained later in this chapter).
- Devices damaged from a drop may have obvious damage such as a cracked screen, or they may have hidden damage. See the sections "Taking It Apart" and "Components with a Hairline Crack" later in this chapter.

General Repairs and Troubleshooting | 153

First Things to Check

Before getting out the tools and taking the device apart, perform a few basic checks first.

Is It Getting Power?

Check to make sure that the power cord is firmly plugged into the device and the wall. If there's an outlet strip involved, make sure that it is plugged in and turned on. Make sure that the wall outlet has power by plugging something else into it as a test. Sometimes the top outlet is controlled by a wall switch and the bottom one isn't. If an AC adapter is used, make sure that it is the right AC adapter for that device.

For battery-operated devices, make sure that the batteries are good and are installed correctly. Usually, the springs go to the negative battery terminals, but not always. If the terminals are corroded, they may not be making good electrical contact with the batteries. Clean them with a pencil eraser or vinegar and cotton swabs.

Check the Fuse

If there's an easily accessible fuse holder, check the fuse.

Unplug for a Few Minutes

Microprocessors (now in almost everything) can get off in the weeds from power spikes and other conditions and make the device unresponsive. Unplug it for a few minutes; then plug it back in and see if it works.

Factory Reset

More complex electronics such as smart TVs have different levels of reset. When unplugging for a minute doesn't fix a problem, perform a factory reset. Factory reset for TVs is usually located in one of the menus under Settings, System, or Support. Consult the owner's manual or internet for how to perform a factory reset on other devices.

Taking It Apart

Sometimes taking a device apart is a bigger challenge than fixing what's broken. This section gives some suggestions to minimize damage while taking things apart and getting them back together again.

Screws under Feet and Labels

Hiding screws under feet is common. Feet may be adhesive backed or push-in. Usually, screws under feet are an all-or-nothing proposition—you won't find just one.

Another popular place to hide screws is beneath adhesive labels. This acts as a tamper-evident seal for manufacturers. The way to check for these is to run your finger over the entire sticker surface, applying pressure, and feeling for round indentations. If you find one, you have the choice of running an X-Acto knife around the inside of the hole to remove a circle of sticker or using a heat gun (or hair dryer) to warm up the sticker and carefully peeling it back.

> **MAKE YOUR SCREWDRIVERS MORE ATTRACTIVE**
>
> A magnetic screwdriver set saves time and frustration when you are disassembling and reassembling equipment held together with screws. This is especially true when screws are at the bottom of deep access holes. An alternative is to make the screwdrivers you already own magnetic.
>
> The Klein Tools MAG2 Magnetizer (Fig. 8-1) works by swiping the tool shaft through the magnetizing hole in the magnetizer. There are also demagnetizer holes for demagnetizing tools.

FIGURE 8-1 Klein Tools MAG2 Magnetizer.
(Courtesy of Klein Tools.)

Hidden Tabs

Hidden tabs are commonly used to hold two halves of plastic cases together. Examples include remote controls, laptops, routers, and phones. Most devices that use hidden tabs also use screws, so if a case seems like it's not coming apart at a certain spot, there's probably a missed screw somewhere close by. Prying apart cases requires finesse to avoid damage. For phones and tablets, you can buy a pry tool set to make the job easier (Fig. 8-2). If you don't have a pry tool set, credit cards and guitar picks come in handy. There are plenty of YouTube videos showing how to take any popular phone apart.

FIGURE 8-2 iFixit prying and opening tool assortment. *(Courtesy of iFixit.)*

Cutting It Apart

Some plastic cases are ultrasonically welded together and are not made to be taken apart. A prime example is some AC adapters. For such items, you'll need to cut them apart along the seams. Tools that are useful for this job are a mini hacksaw (Fig. 8-3), razor saw (Fig. 8-4), and a hot knife tip for a soldering iron (Fig. 8-5). Care must be taken to avoid cutting too far into the case and damaging internal components. When using a hot

General Repairs and Troubleshooting | 157

knife tip for a soldering iron, turn down the heat. You want to melt the plastic, not burn it.

FIGURE 8-3 Nicholson Little-Nic utility saw. *(Courtesy of Crescent Tools.)*

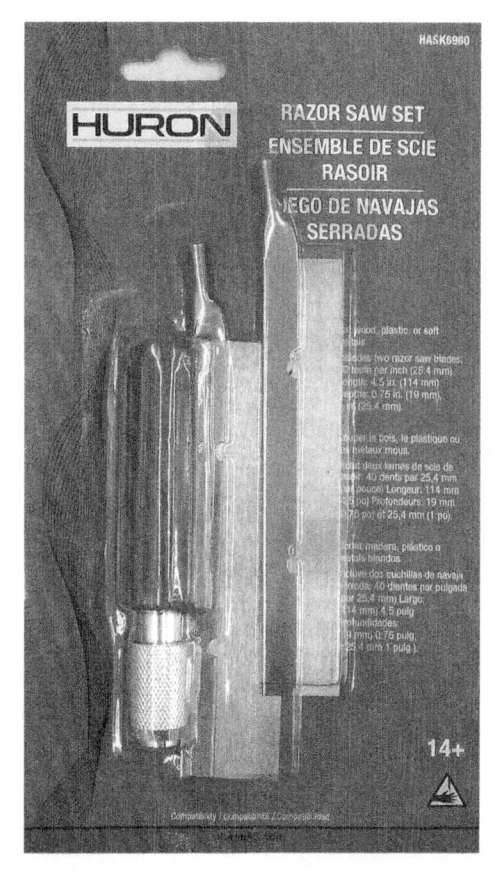

FIGURE 8-4 Razor saw.

FIGURE 8-5 Weller ET Series soldering iron hot knife tip. *(Courtesy of Weller.)*

Keeping Track of Which Screws Go Where

If all the screws that hold a device together are identical, there's no risk of damaging an internal assembly by using a screw that's too long during reassembly. But this rarely seems to be true, so you need a system to keep track of which screws go where.

Egg cartons make great containers for holding screws and other small parts during disassembly (Fig. 8-6). You can fill the compartments in order of disassembly and then reverse the order for reassembly.

I like to use multicolored tape when working on TVs (Fig. 8-7). Most TVs use 20 or more screws of at least two different types to hold the back cover on. The most common screw type gets no tape. But once I come to a screw of a second type, I put a piece of tape next to each screw hole of that type. I also put a piece of tape next to the pile of those screws. When I come to a third screw type, I use a different color tape for those, and so on.

General Repairs and Troubleshooting | 159

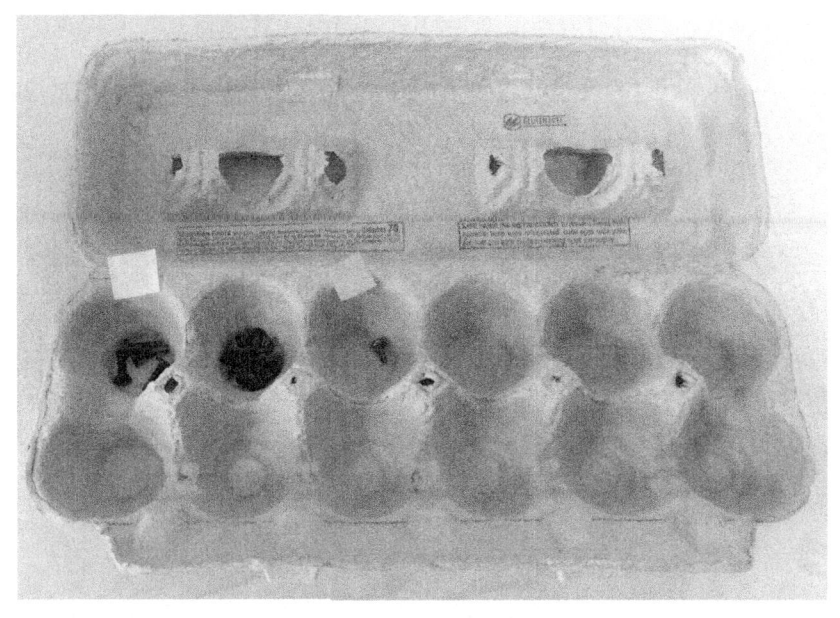

FIGURE 8-6 Egg carton for holding screws and parts.

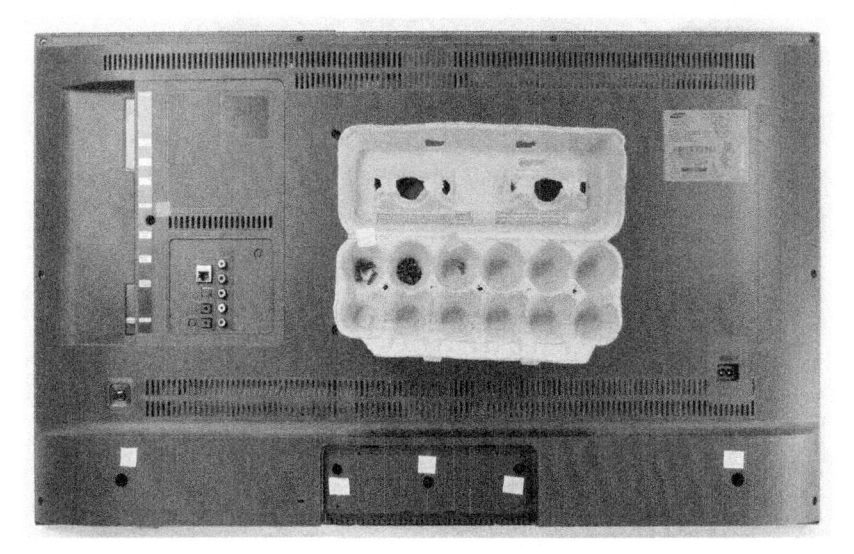

FIGURE 8-7 Multicolored tape labeling screw types on the back of a TV.

Take Pictures

Use your phone to take pictures of the entire disassembly process (Fig. 8-8). This is a great insurance policy for when you're trying to put it back together and can't remember where a screw or other part goes.

FIGURE 8-8 Use your phone to take disassembly photos.

Board Connectors

Electronics are filled with flexible flat cable (FFC) and wire-to-board connectors that need to be disconnected for product disassembly. These need to be taken apart properly to prevent damage. Table 8-1 lists some of the most common types and how to take them apart and put them back together. For connector types not listed, always look for a locking mechanism to release before you force a connector apart and damage the cable or connector.

TABLE 8-1 Board Connector Disassembly

Connector	Take Apart	Put Together
Front flip actuator	Flip the actuator up with your fingernail. Angle the flat cable slightly to allow the two ears to clear the connector, and pull out (Fig. 8-9).	Make sure that the actuator is up. Angle the flat cable slightly to allow the two ears to clear the connector, and insert. Flip the actuator down.
Back flip actuator	Flip the actuator up with your fingernail. Angle the flat cable slightly to allow the two ears to clear the connector, and pull out (Fig. 8-10).	Make sure that the actuator is up. Angle the flat cable slightly to allow the two ears to clear the connector, and insert. Flip the actuator down.
Slider	Slide the plastic retaining bar to its out position by alternately working the two sides of the bar. Pull the flat cable out (Fig. 8-11).	Make sure that the plastic retaining bar is in its out position. Fully insert the flat cable. Slide the plastic retaining bar to its in position.
Non–ZIF	Grasp the flat cable near the connector. Pull the flat cable out (Fig. 8-12).	Grasp the flat cable near the end. Firmly push the flat cable into the connector until it's fully seated.
Tabbed	Squeeze the tab at the back, and pull the cable out (Fig. 8-13).	Insert the cable firmly until the tab engages.

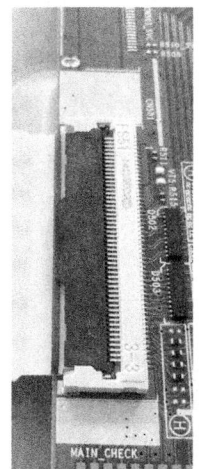

FIGURE 8-9 Releasing front flip actuator connector.

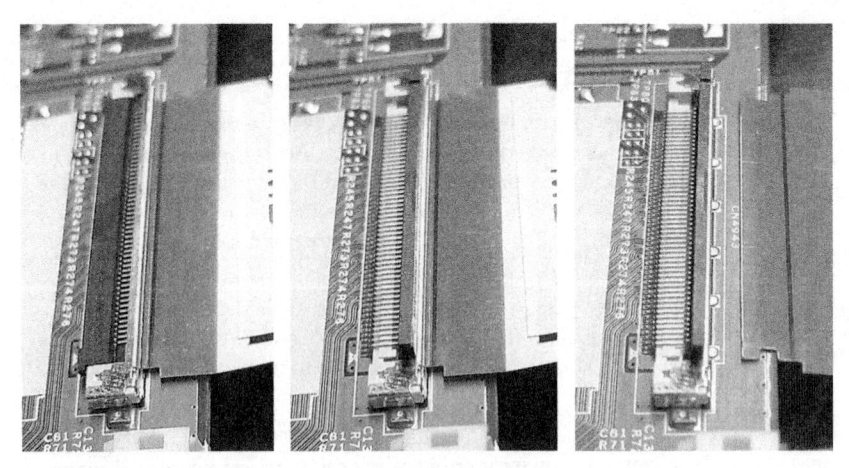

FIGURE 8-10 Releasing back flip actuator connector.

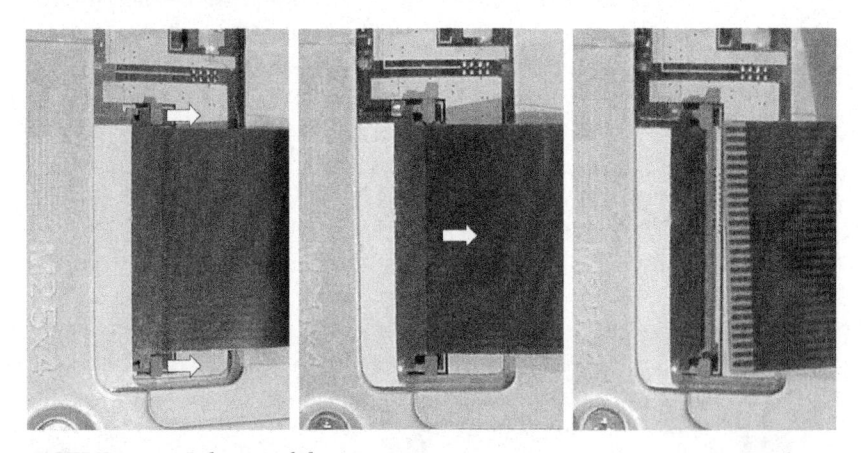

FIGURE 8-11 Releasing slider connector.

General Repairs and Troubleshooting | 163

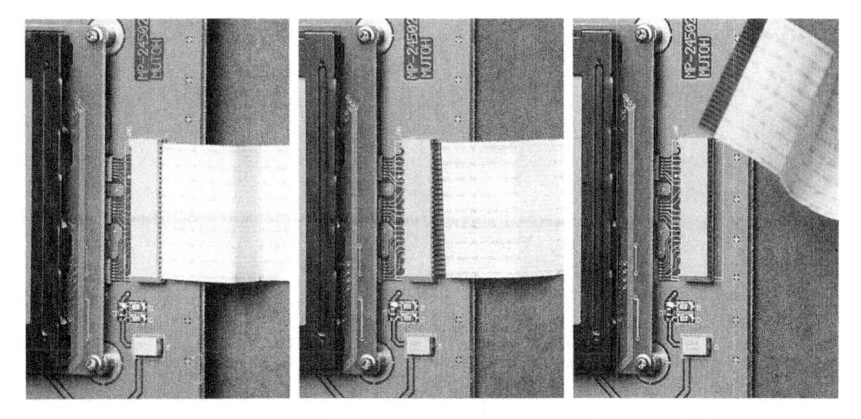

FIGURE 8-12 Releasing non-ZIF connector. *(Courtesy of Chris Purola.)*

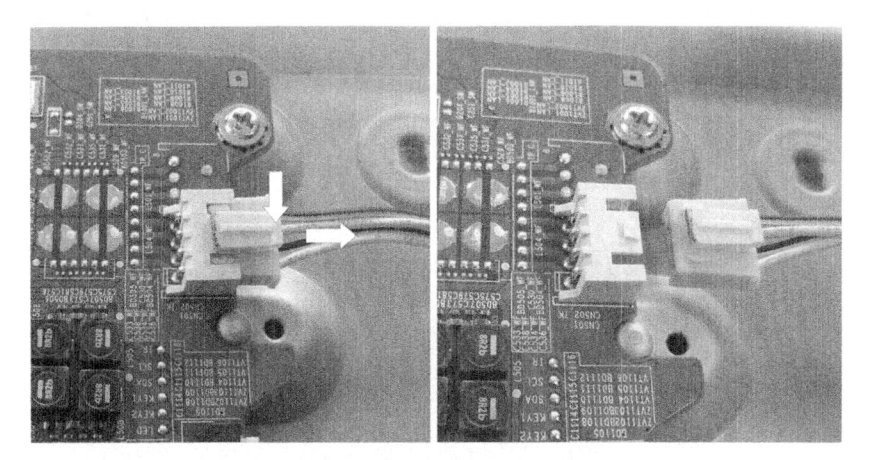

FIGURE 8-13 Releasing tabbed connector.

Visual Inspection

Visual inspection can go a long way toward revealing the source of a problem. This section describes what to look for and what it means.

Bulging or Leaking Electrolytic Capacitors

Failed electrolytic capacitors are a top cause of problems in electronics. Look for electrolytic caps that are bulging, have split tops, or any signs of leakage (Fig. 8-14). If you find any, they're bad. Caps that look fine can also be bad, so if you find a bad one, check others that look good with an ESR or capacitance meter. If the device only has a few electrolytics, consider replacing all of them preemptively. See Chapter 7 for guidance on replacement parts.

FIGURE 8-14 Bulging electrolytic capacitor.

Signs of Liquid or Corrosion Damage

Liquid or corrosion damage could be from a nearby electrolytic capacitor or from condensation or a spill. Some rubber parts liquefy over time and flow. If the area showing signs of liquid or corrosion damage is extensive, the device may not be a good candidate for repair. If it's isolated to a small area, clean it with isopropyl alcohol or water and cotton swabs. Inspect for damage to individual parts, and replace them as needed.

Burned Components

Burned components always have a story behind them, but the story often has a surprise ending. A burned component may be the original cause of a problem or the victim of another part failure that caused the burned part to fail with it. Sometimes the heat from one component can scorch one of its neighbors just by physical proximity.

Always check components that are functionally related to the burned one or you may end up burning the replacement when you turn the power on. It helps to have some knowledge of the circuit to find related parts, but if not, you can follow the PC board traces from the burned part to other good candidates to check. Suspect transistors when a resistor is burned. Check other transistors and resistors when a transistor is burned.

If the PC board is scorched, clean the area with isopropyl alcohol, and try to scrape off any charred areas of the board because these may be slightly conductive. Repair damaged PC board traces as necessary (see Chapter 5 for guidance).

Deformed Components

Deformed components such as bloated power resistors can result from sustained overheating. Treat deformed components like burned ones, looking for causes of failure and associated damage in related circuitry.

Melted Insulation on Wires

Wires with melted or burned insulation are usually the result of a short circuit involving the wire itself or the circuit connected to the wire. Look for burned or damaged components nearby because the wire is usually the victim of a failure elsewhere.

Components with a Hairline Crack

Spotting hairline cracks requires a careful inspection under magnification. These tiny cracks can be found on resistors, transistors, diodes, ICs, and surface-mount components of all types. Use a dental tool or other small probe to push on the leads or the body of parts to make the crack widen and see if it's really a crack. To tell if a suspected crack in a chip part is real, you may need to use a soldering iron to melt the solder on one side of the part to see if the chip separates. Cracks can be caused by overheating of the part or from stress on the PC board.

Bad Solder Joints

Good solder joints can go bad as a result of the mechanical stress from repeated heating and cooling cycles of the electronics from normal use. The most common locations for bad solder joints are through-hole connector pins, through-hole IC pins, and power resistors. The solder joints of PC board mounted user controls such as a volume-level potentiometer can be subjected to mechanical stress from bumps to the control knob. The problems caused by bad solder joints are often intermittent.

Spotting bad solder requires a careful inspection under magnification and somewhat of a trained eye. Resolder any suspicious solder joints along with their companions on the same connector or component.

Components with a Chunk Missing

Sometimes a part will explode when it fails. You may even find component pieces inside the case. Treat exploded components like burned ones, looking for causes of failure and associated damage in related circuitry.

Discolored Areas on the PC Board or Cabinet

Darkened areas on the PC board or interior of the cabinet can be the result of years of moderate heat from a nearby component such as a power resistor or the result of just minutes of high heat. If there are signs of smoke, it's the latter. In either case, the heat-generating part is a good failure candidate, so check it.

Likely Suspects

Some parts have a much poorer track record than others when it comes to reliability. This section points you to the likely suspects for failure.

Electrolytic and Tantalum Capacitors

Electrolytic capacitors top the list of likely failures. They often show visible signs of failure, such as bulging or split tops or leakage. Often there's no visible sign. Tantalum caps are prone to failure, usually as a short circuit on the power supply, but they're used infrequently compared with electrolytics.

High-Power Components

Components that are directly subjected to high power or are in high-power circuits are at a higher risk of failure. They're subjected to higher voltages or currents that electrically stress the parts and thermal cycling that mechanically stresses them. High-power components include power transistors, power supply diodes, varistors, and any heat-sinked parts.

> **STRESS AND THE COEFFICIENT OF THERMAL EXPANSION**
>
> The reason temperature cycles cause mechanical stress failures in electronics is the coefficient of thermal expansion. Most materials expand as they get warmer, but the coefficient of thermal expansion is different for different materials. This means that as the temperature changes, some things expand more than others. A prime example is connectors and PC boards.
>
> If a 20-pin connector and the section of a PC board to which it mounts are the same size at room temperature, they may want to expand to different sizes after the device has warmed up. This creates stress that over thousands of temperature cycles causes a connection failure at a solder joint. The effect is magnified with larger components such as connectors, ICs, metal shields, and heat sinks.
>
> To minimize these problems, diligent circuit designers try to specify PC boards and components that have low or matching coefficients of thermal expansion.

I/O Components

I/O components are the input and output connectors and their associated electronic components. They're subjected to physical abuse whenever something is plugged into or removed from them. They're also subjected to ESD from the people who plug things in.

If a problem is related to an input or output feature, inspect closely for broken connections at the associated connector pads and cracked components nearby.

Mechanical Components

Mechanical components such as switches, relays, pots, and connectors are frequent causes of problems because they are sensitive to dirt, contamination, moisture, wear, electrical arcing, and oxidation. Contact cleaners can often renew switches, connectors, and pots. Reseating connectors (taking them apart and putting them back together) can fix a bad or intermittent connection.

Intermittent Problems

Intermittent problems come and go. They're especially challenging to fix because you can't find the problem when the device is working properly.

Get the Problem to Show Up

The first step in fixing intermittent problems is trying to get the problem to reliably show up. This gives you a way to work on the problem and a way to know when it's fixed. It also provides clues about what's broken. Observe the conditions that make the problem appear, such as the temperature of the room, how long the device has been on, mode of operation, and so on.

Here are some things to try with the cover off and the product operating. Use caution. If working near potentially lethal voltages, use an isolation transformer, and keep one hand behind your back.

- Wiggle wires, cables, and connectors.
- Try different product orientations where appropriate.
- Press on the circuit board in various locations using an insulated object such as a plastic pen.
- Use a heat gun to selectively heat components.
- Use freeze mist to selectively freeze components.

Likely Suspects for Intermittent Problems

Table 8-2 shows some likely suspects for intermittent problems and recommended actions.

TABLE 8-2 Likely Suspects and Actions for Intermittent Problems

Likely Suspect	Action
Connectors	Reseat (take apart and put back together)
Socketed ics	Reseat (remove and reinsert)
Switches and pots	Spray with contact cleaner and then work the part, resolder leads if PC board mounted
Relays	Tap with screwdriver handle when energized to see if it switches
Bad solder joints	Resolder all suspicious joints
Hairline crack in component	Careful inspection under magnification
Failing electrolytic capacitor	Test with ESR meter or parallel-cap technique

When inspecting for and resoldering suspicious joints, resolder any joint that looks dull or cracked. Resolder any joint where the component lead may not have been pushed far enough through the hole when it was originally soldered. This can happen in manufacturing, and the product can work properly for years before the unsoldered mechanical contact fails. Push the component lead further through the hole when resoldering.

In-Circuit Component Testing

In-Circuit versus Out-of-Circuit Testing

In many cases, the only way to know for sure whether a component is good or bad is to test it out of circuit. This is because in-circuit readings are affected by other components. Unfortunately, removing components from the PC board is time-consuming and risks damaging the PC board. This section provides a strategy for in-circuit testing to reduce time and effort. Unless stated otherwise, the procedures in this chapter assume the circuit is not powered.

Measure In-Circuit

Measuring components in-circuit can provide useful information if interpreted correctly (Table 8-3). Take resistors as an example. If a 100-ohm resistor measures 100 ohms in-circuit, this probably means that the circuit isn't affecting the ohmmeter reading and the resistor is good. If, however, the 100-ohm resistor measures 50 ohms, the result is inconclusive. This is because the circuitry could be reducing the reading. If the 100-ohm resistor measures a steady 200 ohms, this means that the resistor is bad. This is because other components could only reduce the reading, not increase it. (Note that a charged capacitor can cause incorrectly high ohmmeter readings, but they'll change as the cap charges or discharges.) Reverse the ohmmeter leads and measure the resistor again. If the readings are different but stable, consider the higher one as the better reading.

TABLE 8-3 In-circuit Component Measurement

Component	Comment
Resistors and potentiometers	Resistance too high means bad part; too low may be okay
Electrolytic capacitors	ESR meter works in-circuit Capacitance meter: too low means bad; too high may be okay
Diodes	Diode test over limit (OL) for both polarities means bad part
Transistors	Bipolar transistor test may work in high-impedance circuits
Fuses	Okay to check continuity in-circuit

Testing diodes with the diode test feature of a multimeter should show over limit (OL) for one polarity and about 0.6 volt for the other. Like resistors, other components serve only to reduce the readings. If both test polarities show OL, the diode is open circuited.

In-circuit transistor testing may be possible for bipolar transistors (NPN or PNP) in high-impedance circuits. Use the out-of-circuit methods (in Chapter 7) to perform in-circuit testing. If the transistor fails the test, you'll need to check it out of circuit to know for sure.

Lift One Lead

A shortcut to fully removing a through-hole component for testing is to desolder one of the leads, leaving the other lead soldered to the PC board (Fig. 8-15). This electrically removes it from the circuit. For components with more than two leads, you'll need to lift all but one lead.

FIGURE 8-15 Component with lifted lead.

If you can see the PC board traces (or have a circuit diagram), you can lift the lead of a series-connected component instead. The series-connected component may be easier to access or may be a through-hole part that is series connected to a surface-mount part you want to electrically isolate. In Figure 8-16, R2 can be measured in-circuit because series-connected R1 has a lifted lead.

FIGURE 8-16 Series-connected component with lifted lead.

The Joy of Sets

Sometimes the circuit you're working on has a duplicate you can use as a reference. Common examples are audio equipment with left and right channels and dual power supplies. Or you may have access to multiple units of the same model, one working and one not. This gives you the ability to perform in-circuit testing to a higher level. It no longer matters if a resistor's reading is affected by other circuit components; it should measure the same as its twin.

When using this technique,

- The polarity of your multimeter leads must be the same for both measurements.
- Both channels must have the same settings and loads.
- Differences may not be from the component you're measuring.

Out of circuit resistor measurements are insensitive to which side the red or black probes are connected to, but that's not true for in-circuit measurements, where there could be a diode or transistor connected to the resistor.

All the settings and external connections must be the same for both measurements or you'll be comparing apples to oranges. For stereo receivers, set the balance control to the center position.

If you measure a component difference between the bad and good channels, keep in mind that it might not be from the part you're measuring. It might be due to another part that's affecting your measurement. Measure the suspect component out of circuit or perform additional measurements on nearby components to determine the culprit.

Power the Circuit

Applying power to the circuit allows better in-circuit testing of transistors, Zener diodes, voltage regulators, and transformers (Table 8-4).

TABLE 8-4 Powered Component Measurement

Component	Comment
Bipolar transistors	Check for greater than 0.7 volts between base and emitter and for the same voltage on any two terminals.
Zener diodes	In voltage references and regulators, DC voltage across diode should be close to the Zener voltage.
Voltage regulators	See Chapter 7.
Transformers	See Chapter 7.

SAFETY TIP

Never work on powered circuits unless you know and use safe working practices. Many circuits that derive power from 120 VAC, and some that don't, contain lethal voltages as well as other hazards. If you are working near potentially lethal voltages, use an isolation transformer, and keep one hand behind your back.

Bipolar Transistors

Powering a device often lets you detect transistor failures in-circuit. Testing transistor circuits is more advanced and is covered in the section "Troubleshooting." The following test is for bipolar (NPN or PNP) transistors. Use your multimeter to measure the DC voltages of the three transistor terminals.

- More than an approximately 0.7-volt difference between base and emitter voltages may indicate an open-circuit base-emitter junction.
- The same voltage on two or more terminals may indicate shorted junctions.

While the same voltage on multiple terminals can indicate a bad transistor, it can also be caused by other circuit conditions.

Zener Diodes

Zener diodes are useful for voltage references, power supply regulators, and overvoltage protection circuits. Zeners usually fail as a short, which is easily detected in-circuit with the diode test feature of a multimeter. To see if a Zener diode is working properly in a voltage reference or power supply regulator circuit, measure the DC voltage across it. It should be approximately equal to its specified Zener voltage. For overvoltage protection circuits, the DC voltage should be less than the Zener voltage.

PARALLEL-CAP TECHNIQUE

If you don't have an ESR meter or capacitance meter, detecting bad electrolytic caps is difficult if there are no visible signs of failure such as a bulging top or leakage.

One option is to measure the AC ripple voltage across each cap with the circuit powered, but you may not know what an acceptable AC level is. Another option is the parallel-cap technique (Fig. 8-17). The idea is to temporarily connect a good capacitor across the suspect capacitor to see if the problem goes away. Connect the cap using clip leads or by tack soldering. (Tack soldering just means a temporary connection to hold a component in place.) The power should be off while making the connection. When using this technique, follow these rules:

- The parallel-cap polarity must match the suspect cap (get it wrong and it could explode).
- The parallel-cap voltage rating must equal or exceed that of the suspect cap.
- The parallel-cap capacitance should be about the same as that of the suspect cap.
- For capacitances larger than 100 µF, safely discharge the suspect cap before connecting the parallel cap.

It's handy to have a stockpile of higher-voltage electrolytic caps of various values for this technique. If the parallel capacitance value is less than about half that of the suspect cap, it may not be enough to affect the problem. If it's more than the original, it could cause circuit malfunctions.

FIGURE 8-17   Parallel-cap technique.

More Joy of Sets

With the power turned on, having a duplicate circuit for reference provides even more clues about what's broken. But aimlessly measuring the AC and DC voltages of every component to compare against its twin is not a good strategy. Use this technique to measure and compare the voltages of highly suspected parts. Then it's time to move on to troubleshooting.

Troubleshooting

Up to this point in this book, the focus has been on detecting component failure. The advantage of this approach is that it requires little or no understanding of how a circuit or system works. For more complex problems, though, using component failure detection is like duck hunting blindfolded.

What Is Troubleshooting?

Troubleshooting is performing a series of experiments based on knowledge of how a circuit or system works. The goal is to find and fix the problem. Entire books have been written on troubleshooting electronics. I'll try to condense the subject to its essence.

Before troubleshooting, start every repair project by checking the obvious and easy stuff first. This includes the component failure detection methods described previously.

Understand the System

If you're going to be performing experiments based on knowledge of how a circuit or system works, the first step is to gain that knowledge. Table 8-5 shows some top resources.

TABLE 8-5 Resources for Understanding the System

Resource	Comments
Service manual	Single best resource when you can find it
IC datasheet	Useful when problem centers around an IC
YouTube videos	Great for common problems on popular devices
Repair forums	Can ask questions
iFixit.com	One-stop online repair resource

The internet is where you'll find most resources to improve your knowledge of a particular device. When it's available, the single most useful resource is a service manual. Comprehensive service manuals contain descriptions of how a device works, circuit diagrams, parts information, and troubleshooting guidance. Sometimes a service manual is available for free downloading; other times you'll have to pay for an electronic or paper manual. If you can't find a service manual for your exact product, try searching for a related model. Often the related product manual is similar enough to be useful. At other times, service manuals cover multiple products, and you'll only find the manual you want by searching for the related product.

When a service manual isn't available, another good resource is IC datasheets. These are produced by IC manufacturers to guide circuit designers and often include example application circuits that resemble what's in your product. Look for datasheets of ICs in the suspected problem areas of the device you're fixing. Include the word *datasheet* in your internet search along with the part number of the IC.

YouTube is an amazing source of repair videos for almost anything. Some videos are more professionally done than others, so don't take everything you see as fact. As with service manuals, search for related models if you're not finding what you want. I have the highest respect for those who take the time to make and post repair videos. If you succeed at fixing your device and didn't find a good repair video, consider giving back and posting one.

Repair forums are another excellent online repair resource. A Google search on your item along with terms such as *repair* and *fix* will often turn up links to repair forums. If you're not getting good search results, you may need to go to a specific forum website and perform your search there. Once on the forum, you can post questions, and members will try to help. Keep in mind that members are most interested in answering well-thought-out, specific questions.

iFixit.com is positioning itself to be a one-stop online repair resource. The company is a big advocate of the "Right to Repair" movement. The site provides repair guides for electronic and nonelectronic items, product teardowns, and its own repair forum. iFixit also supplies specialty repair tools and parts.

Draw Your Own Partial Schematic

A schematic (circuit diagram) goes a long way toward understanding a circuit. If you can't find one for your product, draw your own. It doesn't need to be complete; it only needs to cover the part of the circuit you're interested in. Follow the PC board traces as best you can, and show unidentifiable components as boxes with wires. Try to identify ground and power lines on your circuit board and show them using ground and power symbols in your diagram. A continuity tester helps to confirm accuracy. Often redrawing the diagram to clean it up helps show the circuit structure better and make its function clearer.

Divide and Conquer

Once you have some knowledge of how your device circuitry should work, the second part of troubleshooting is devising experiments to home in on the problem. The most effective strategy is "divide and conquer." The ancient Romans may not have repaired electronics, but they understood the effectiveness of *divide et impera*. For electronics, this means repeatedly dividing the suspected problem area in half and using the process of elimination until you've found the problem.

Using a stereo receiver with no sound at the speakers as an example (Fig. 8-18), you might start by first dividing the problem into the power supply circuit and the non–power supply circuit. You would devise an experiment to check the power supply circuit by measuring all the power supply voltages. If everything looked good with the power supply, you would next need to divide the non–power supply circuit in half. You might divide it into the power amplifier circuit and the non–power amplifier circuit. You might then devise an experiment to check the non–power amp circuit by selecting the FM radio mode, hooking up an antenna, and using an oscilloscope to check for an audio signal at the input to the power amp stage. If you found no signal at the power amp input stage, the next step would be dividing the non–power amp stage in half and looking for signal at the halfway point. The process continues until you've found the point where the signal stops.

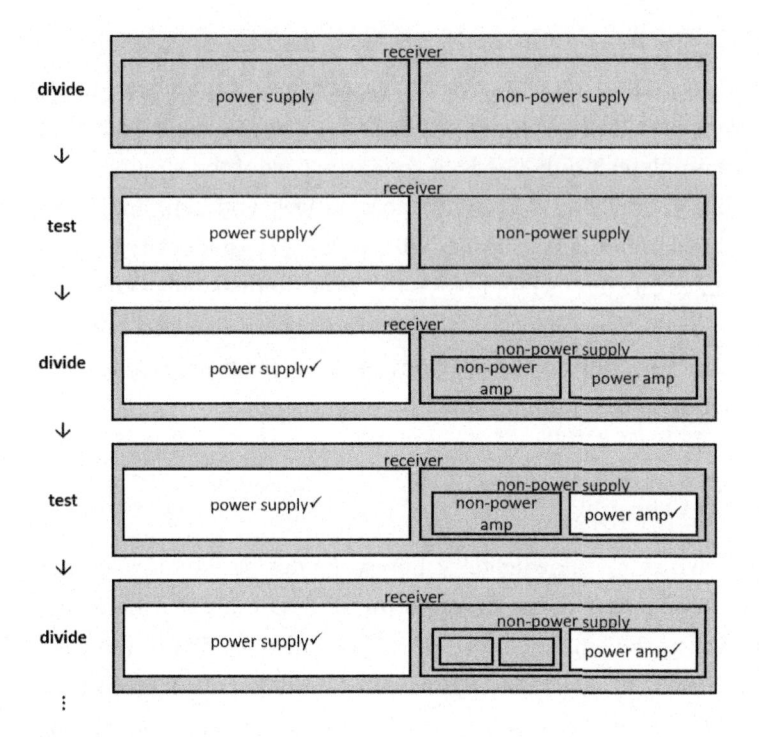

FIGURE 8-18 Stereo receiver divide and conquer.

Fix the Cause

Troubleshooting requires knowing something about how a circuit or system works. This helps you to identify the cause of damage, not just the result. Never assume that a component broke for no reason. When you find a failed part, always ask what other components could be damaged by the failure and what damaged parts could have caused the failure.

Power Supply Troubleshooting

Checking the power supply in a malfunctioning device is usually a good first step in dividing and conquering. For this section, a power supply could be an AC adapter or it could be power supply circuitry inside the main product enclosure. Some power supplies are derived from other power supplies, and these would be included too. For example, a 5- or

3.3-volt logic supply for a microprocessor might come from a three-terminal voltage regulator using as its input the 12-volt supply provided by an AC adapter.

BASIC REGULATED POWER SUPPLY CIRCUIT

Figure 8-19 shows a basic circuit for a regulated power supply. This circuit, or some variation of it, is commonly used, so it's useful to understand its operation. A transformer converts the 120-VAC input to a lower AC voltage at the output of the transformer's secondary winding. A diode bridge composed of four diodes converts the AC voltage to rectified AC. Electrolytic capacitor C1 smooths the rectified AC to produce high-ripple DC. The LM7812 voltage regulator produces regulated 12 VDC at the circuit output. The purpose of capacitors C2 and C3 is to improve the transient response of the regulator.

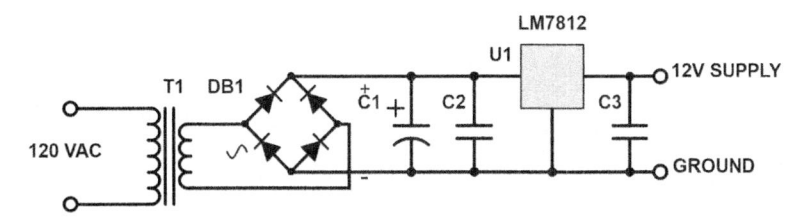

FIGURE 8-19 Basic regulated power supply circuit.

When a power supply voltage is incorrect (low, high, or too much AC ripple), you'll need to determine whether the power supply, the load circuit, or both are the problem. (The load circuit is the circuit that uses that supply for power.) A failure in a load circuit can damage a power supply. To isolate the problem, you'll need to find a way to disconnect the load circuit from the output of the power supply. Unless the power supply is a separate PC board (or an AC adapter), this usually means desoldering a component lead or two. If the power supply voltage looks correct with the load disconnected, it's time to look for a problem in the load circuit, such as a shorted transistor or tantalum capacitor.

Most ESR meters provide high-resolution low-resistance measurements even when not connected to a capacitor. This capability can be used to home in on load circuit power supply shorts. Connect one probe to ground and use the other probe to navigate the shorted power supply traces on the PC board to find the location of the minimum resistance value. That's where the short is.

Another method to find power supply shorts is *current blasting*. The idea is to connect a high-current external power supply to the disconnected load circuit and watch for signs of heat as you turn up the supply current. Set the supply to the nominal voltage and set its current limiter adjustment to 0 amps. Turn up the current gradually, watching for smoke or a glow from the shorted component. Back down the current as soon as you see the culprit to avoid unnecessary damage to the PC board. Wear safety glasses and exercise caution when performing this procedure.

Transistor Circuitry Troubleshooting

Troubleshooting transistor circuits is best done with an oscilloscope, with which you can see the actual waveforms at each of the terminals, not just measure DC voltages.

Bipolar Transistor Amplifiers

If a bipolar (NPN or PNP) transistor is being used in an amplifier, use these simplifying troubleshooting guidelines:

1. The base is a high-impedance input.
2. The DC voltage at the emitter is approximately 0.6 VDC lower than the base for NPN (0.6 VDC higher than the base for PNP).
3. The AC waveform at the collector is an inverted and amplified version of that at the base (common emitter) or the AC waveform at the emitter is the same as that at the base and the collector has no AC (common collector).

Check the DC voltages at the base and emitter for the expected DC offset. If the AC waveforms correspond to one of the conditions described by the third assumption, the transistor should be assumed to be good. The rarely used common-base topology has no AC signal at the base or emitter but an output signal at the collector.

MOSFET Amplifiers

If an enhancement-mode MOSFET is being used in an amplifier, use these simplifying guidelines when troubleshooting:

1. The gate is a high-impedance input.
2. The DC voltage at the source is a few volts DC lower than the gate for n-channel (a few volts DC higher than the gate for p-channel).
3. The AC waveform at the drain is an inverted and amplified version of that at the gate (common source) or the AC waveform at the source is the same as that at the gate and the drain has no AC (common drain).

Check the DC voltages at the gate and source for the expected DC offset. If the AC waveforms correspond to one of the conditions described by the third assumption, the transistor should be assumed to be good. The rarely used common-gate topology has no AC signal at the gate or source but an output signal at the drain.

Transistor Switching Circuits

When transistors are used as switches (to energize a relay coil, for example), they are either fully off or fully on. Table 8-6 shows switching circuits for the four most common transistor types. In practice, the p-channel enhancement-mode MOSFET is a distant fourth because it's rarely used. In the bipolar circuits (NPN or PNP), the base resistors perform the necessary job of limiting the base current to prevent destroying the transistor. In MOSFET circuits, the gate resistor mainly reduces ringing and is often omitted. In all cases, when the switch is off, R_{load} has 0 volts across it, and when the switch is on, R_{load} has Vcc across it.

Troubleshooting transistor switching circuits is a matter of observing and comparing the waveforms at the input and output of the switch.

TABLE 8-6 Transistor Switching Circuits

Off	On	Circuit
Vin ≈ 0	Vin > 1 volt	NPN transistor switching circuit: VIN through Rb to BASE; COLLECTOR connected via Rload to Vcc; EMITTER to ground.
Vin ≈ Vcc	Vin < Vcc −1 volt	PNP transistor switching circuit: EMITTER to Vcc; VIN through Rb to BASE; COLLECTOR through Rload to ground.

Off	On	Circuit
Vin ≈ 0	Vin > 3 volts	N-CH MOSFET with Rload from Vcc to Drain, Source to ground, VIN through Rg to Gate
Vin ≈ Vcc	Vin < Vcc −3 volts	P-CH MOSFET with Source to Vcc, Drain to Rload to ground, VIN through Rg to Gate

BASIC TRANSISTOR AMPLIFIER CIRCUIT

Figure 8-20 shows a common-emitter amplifier circuit. There are many transistor amplifier configurations. This is one of the most popular and shows how voltage gain is achieved with transistors.

Operating Regions

Bipolar transistors have three regions of operation: cutoff, active, and saturation. Cutoff and saturation can be considered fully off and fully on, respectively, and are useful when using transistors as switches. For amplifiers, the active region is used.

How Does It Work?

The purpose of resistors R1 and R2 is to set the DC voltage at the base of the transistor. In an amplifier, the DC operating point is optimized to provide the greatest undistorted output range. The purpose of capacitor C1 is to block the DC voltage from the amplifier input so that R1 and R2 can do their job.

Here's how amplification occurs in this circuit. Note that both the input and output voltages are relative to ground. The input AC waveform Vin is coupled with the base through C1. With emitter loading as shown and in the active region, the AC waveform at the emitter is the same as the base.

Here's where the amplification magic happens. The current through the collector is approximately the same as the current through the emitter. Using Ohm's law,

$$\text{Emitter AC current} = \text{Vin}/\text{Re}$$
$$\text{Collector AC current} = -\text{Vout}/\text{Rc}$$

Setting these currents equal and doing the math,

$$\text{Vout} = -\text{Rc}/\text{Re} \times \text{Vin}$$

This shows that the voltage gain is the ratio of −Rc/Re. If Rc is five times the value of Re, the amplifier would have a voltage gain of −5, and the AC output would be 500 mV for a 100-mV input. The minus sign means that the output waveform is inverted (upside down) with respect to the input.

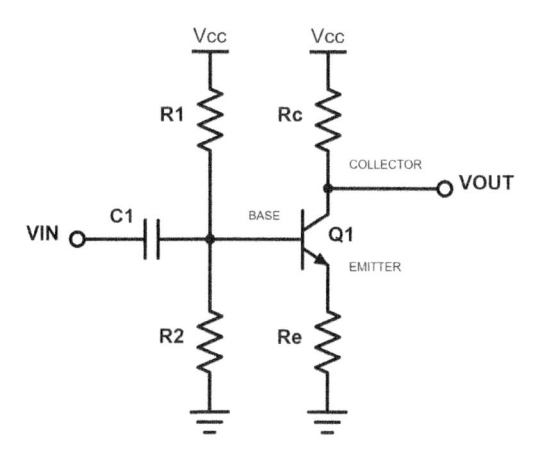

FIGURE 8-20 Basic transistor amplifier circuit.

Keeping a Repair Log

At some point, I realized that many of my new repairs were duplicates of past repairs. I would sometimes rediagnose a problem I had fixed before, not remembering the fix. That's when I decided to keep a repair log.

Keeping a repair log can be as simple as you want it to be. You can use a spiral notebook or Excel spreadsheet. It's a small investment in time that can pay back in a single duplicate repair. An excerpt of my repair log is shown in Table 8-6.

TABLE 8-6 Repair Log Example

Date	Owner	Type	Model	Problem	Parts	Notes
x/x/xx	x	dsl modem	2wire 2701hg-b	dead	1000µF 16v 105c low esr 470µF 10v 105c	Lytics inside wall wart were bulging and measured near zero.
x/x/xx	x	tv	insignia ns-lcd37-09 ver a	dead	audio video main board CBPF8Z-5KQ1 715T2830-1	LED turned blue and supplies came up, but no pix, sound or backlight.
x/x/xx	x	speakers	bose 501 series iv	bad woofers	bose 111791 replaced with Parts Express 295-315	damaged voice coil. Replaced woofers with Dayton DC250-8 10".
x/x/xx	x	carpiagiani cpu module		level sensor error		Bad solder on pin.
x/x/xx	x	computer	imac g5	intermittent won't turn on	9 lytics in power supply	2 caps vented, 4 bulging, 3 meas bad. See Mouser order of x/x/xx for replacement parts.
x/x/xx	x	ipod docking alarm clock	ihome ih5	right selector knob flakey		pop off knob, spray with contact cleaner
x/x/xx	x	tv	vizio sv470xv1a	no backlight	6632L-0487A backlight inverter slave	Unplugging bad slave board allowed 24v supply to come up on master board. Transistors on slave shorted.
x/x/xx	x	integrated amp	jvc ja-s55	phono preamp bad channel	x101 2sk240v/bl dual fet	adjust bias via r113
x/x/xx	x	jeep module	1994	bad caps	3-220µF 25v	jell encapsulated
x/x/xx	x	high resistance meter	keithley 6517a	overcurrent error, v source stuck at −145v	Q314 [2N7000], Q313 [2SK1412] and Q317 [2SK1412]	The 2SK1412's didn't fail as pure open or short
x/x/xx	x	audio receiver	panasonic SA-BX500PP	dead	TDA8920BJ amp IC	originally suspected power supply and replaced switch mode reg

General Repairs and Troubleshooting | 189

Date	Owner	Type	Model	Problem	Parts	Notes
x/x/xx	x	ups	apc be750g	beeps	c7 330µF 25v, ry1/ry2 omron g2rl-24 12vdc	cap bulging, relay stuck.
x/x/xx	x	tv	sharp lc32sb24u	locked up, error e203 no broadcast now		perform factory reset
x/x/xx	x	subwoofer	velodyne dps-12	dead	2a slow blow fuse	replacing fuse requires pulling plate amp out.
x/x/xx	x	audio receiver	sony str-de185	intermittent right speaker		Resoldered all pins of all relays.
x/x/xx	x	audio receiver	rca rt2400r	no r audio for radio, intermittent r audio for jacks, unit sometimes resets.		Resoldered all three-pin regulators on main PCB. Easy access through removable bottom plate.
x/x/xx	x	broaster cpu module	ncc solid state fryer control tbc-fc440-010	burnt relay contacts	omron g6b-1114p-us 5vdc	1174p has same footprint but slightly taller and increases current from 5A to 8A
x/x/xx	x	audio receiver	technics sa-r230	dead except for lit standby LED and eq LEDs. Before that, had been shutting down after shorter and shorter warmups.		Bad solder on all three pins of Q708 negative supply regulator. Used service data of sa-r330.
x/x/xx	x	broaster cpu module	ncc solid state fryer control tbc-fc445-010	module cycles off & on every few seconds	C4 10µF 35V	Symptoms seemed to subside after warm-up. Freeze mist was used to isolate problem to surface-mount lytic C4.

(continued on next page)

TABLE 8-6 Caption needed (*continued*)

Date	Owner	Type	Model	Problem	Parts	Notes
x/x/xx	x	audio receiver	Denon AVR-1908	shuts down after 2 seconds	KIA7815API 15V regulator (NTE1972)	All power supplies measured during 2 second powerup. 15V and 5V supplies down. 5V reg not replaced as 5V derived from 15V and 5V regulator tested OK.
x/x/xx	x	power supply	hp 6234a	v1 output goes into current limiting when set above a fraction of a volt	CR2 1N5059	Output reverse protection diode failed as a low impedance.
x/x/xx	x	hypot jr high voltage power supply	4030	current meter reads low, voltage range switch intermittent	ammeter bridge rectifier VE67 (NTE5305), toggle switch	
x/x/xx	x	dvd/vcr	Toshiba dvr670ku	all lights blink on and off	C1017 4700µF 6.3V	Lytic was vented. Part is located on power supply board beneath dvd drive.
x/x/xx	x	tv	samsung ln40a550p-3fxza	dark picture, contouring and solarization	AS19H1G	Bad AS19H1G buffer IC on T-CON board V400H1-C03 35-D026047.
x/x/xx	x	audio receiver	pioneer sx312r	phono preamp bad channel	Q2 2SC4689 Q4 2SA1804 R552 100 ohms R556 100 ohms R625 100 ohms D754 1N4148	Used SX311R service manual. NTE37MCP contains matched complementary pair Q2/Q4.
x/x/xx	x	power supply	bk precision 1651	low current limit on output A		reseated vcontrol & icontrol pot connectors on main board.
x/x/xx	x	broaster cpu module	ncc solid state fryer control tbc-fc445-010	module powers up after many minutes or not at all	C41 1µF 50V	Capacitor in parallel technique for all lytics was used to isolate problem to C41.
x/x/xx	x	scope	hp 54600a	vertical height reduced and nonlinear	c608 and c609	10µF 16v and 1000µF 16v on CRT PCB

CHAPTER 9

Product-Specific Repairs

This chapter provides specific repair advice for some of the most commonly repaired products, including flat-screen TVs, laptops, rechargeable battery packs, audio receivers, and remote controls.

The purpose of this chapter is not to cover every problem for every device but to give an idea of the scope of common repairs and provide guidance if no model-specific resources are available. Online resources such as iFixit.com have model-specific repair guides for thousands of devices and are a good place to start when considering any repair.

Flat-Screen TVs

Few products have changed as radically as televisions have over the years. And TV repair has changed with them. Back in the 1960s, tube testers were commonly found in drugstores, and do-it-yourselfers would bring in shoeboxes of vacuum tubes to check whenever their TVs acted up. In the subsequent decades, transistors and ICs replaced vacuum tubes, and televisions got more complex, pushing television repair beyond the capabilities of most do-it-yourselfers.

Now that televisions have almost a million times the computational power and memory of the Apollo 11 guidance computer, the tables have turned again. TV repair in the age of flat-screen TVs is now largely a matter of swapping broken modules. For those more adventurous, it may include replacing bad electrolytic capacitors in power supply boards. The skill is now in determining which module is bad.

What's Inside a Flat-Screen TV?

Figure 9-1 shows what's inside a typical flat-screen TV. The exact number of modules varies from model to model, but the configuration shown is very common. Table 9-1 lists and describes the typical modules and assemblies.

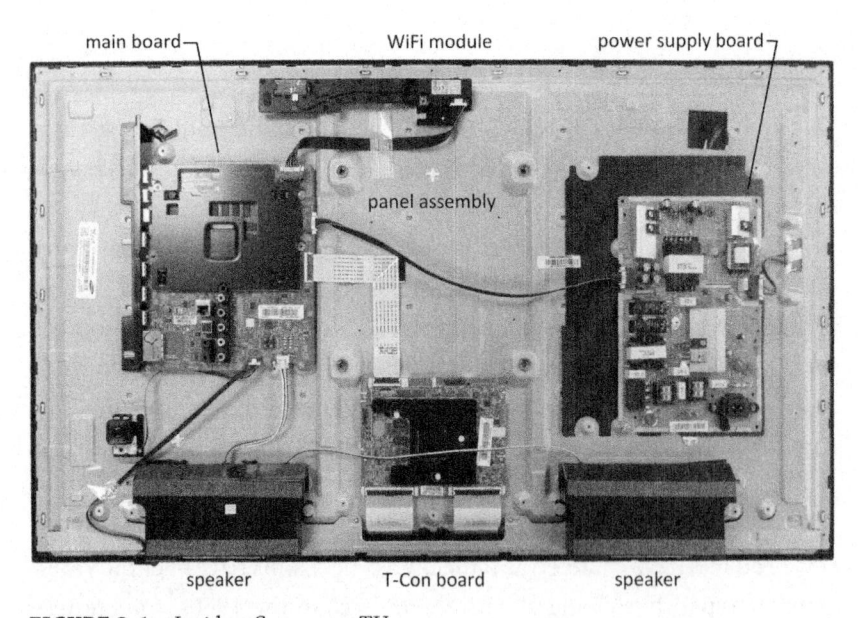

FIGURE 9-1 Inside a flat-screen TV.

Panel Assembly

Panel assemblies generally include a display panel layer as well as a backlight. Liquid-crystal display (LCD) panels do not in themselves emit light and require a backlight behind the LCD layer. In the past, cold-cathode fluorescent lamps (CCFL), miniature versions of fluorescent light tubes, were used for this, but current model LCD TVs use light-emitting diodes (LEDs) for backlighting. These are sometimes advertised as LED TVs, but they are in fact LCD TVs.

Another type of panel assembly, quantum-dot displays, also include an LCD layer but use a combination of blue LEDs and red and green quantum dots to produce a higher-quality backlight than ordinary LEDs.

A third type of panel assembly, organic LED (OLED) displays, emit light at the pixel level, and there is no LCD layer or backlight involved.

Power Supply Board

The power supply board converts the 120-VAC line voltage into the array of DC voltages needed by the other modules. It also provides the often high voltages needed to drive the backlight, which can be in the hundreds of volts for models with LED backlights and 500–700 volts for older LCD TVs with CCFL backlights.

Main Board

The main board performs audio and video processing. The television's microprocessor is also on the main board, generating all the user-interface graphics, responding to remote control commands, and providing smart TV functions.

T-Con Board

The timing controller (T-con) board converts the red-green-blue (RGB) video data and timing information from the main board into the control and data signals required for the row and column driving of the LCD.

TABLE 9-1 Recognizing Modules Inside a Flat-Screen TV

Module/Assembly	How to Recognize	What It Does
Panel assembly	—	Display panel and drive circuitry, includes backlight for LCD displays
Power supply board	AC power cord feeds this board	Provides DC power to other boards and to backlight
Main board	Has signal connectors (HDMI, USB, tuner input)	Performs audio and video processing; includes microprocessor
T-con board	Has three large, flexible flat cable connectors; may be integrated onto main board or panel assembly	Timing controller for the display panel
WiFi/Bluetooth module	Only a few square inches; may be integrated onto main board	Provides wireless internet and Bluetooth capabilities

Taking a Flat-Screen TV Apart

Before taking a TV apart, try a reset, and do a preliminary diagnosis as described in the next section. To disassemble and work on a flat-screen TV, use a flat, sturdy table that's at least as large as the screen. To protect the screen and bezel, cover the table with a blanket before laying the TV screen-side-down on the table. The TV should be unplugged during disassembly and repair.

Remove the TV stand or wall-mount bracket from the back of the TV. Next, remove all the screws holding the back of the TV on. In some cases, the entire back panel doesn't need to be removed; there's a smaller access panel that provides access to the internal modules. Use an egg carton to hold the screws. Use different colors or types of tape on the back of the TV (and the corresponding egg carton compartment) to indicate different screw types. Installing the wrong screw when you put the TV back together can damage internal circuitry if the screw is too long or can strip threads if it's too large in diameter.

Once all the screws are out, gently lift the back off. If there's a spot that doesn't seem to want to lift off, you may have missed a screw. Sometimes there are wires connecting power switches or other assemblies attached to the cabinet back to the main board in the TV, so separate the back from the TV slowly to prevent breaking a wire or connector.

SAMSUNG OPEN JIG

In rare instances, TV manufacturers choose to require special tools for disassembly. For Samsung screwless-back TVs, the tool you'll need is shown in Figure 9-2. There are plenty of online videos demonstrating how to use it.

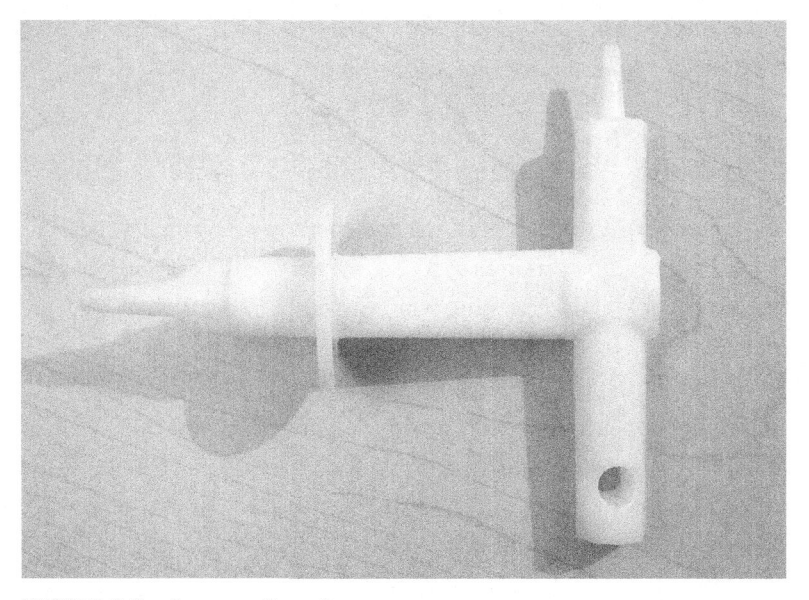

FIGURE 9-2 Samsung Open Jig.

To remove modules for repair or replacement, you'll need to carefully disconnect all cables to that module (Fig. 9-3). Use your phone to take pictures before disconnecting cables if there's any chance of reassembly error. The "Board Connectors" section in Chapter 8 shows the most commonly used connectors in TVs and how to take them apart without damage.

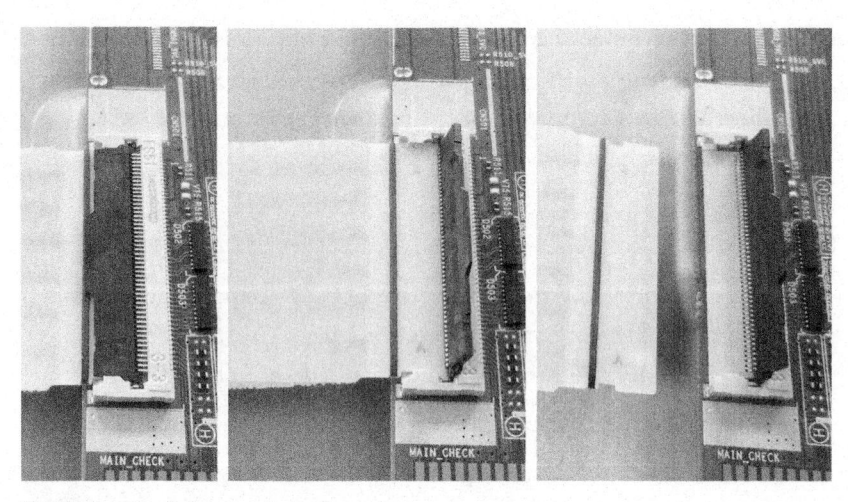

FIGURE 9-3 Releasing front flip actuator connector.

Diagnosing Flat-Screen TV Problems

As mentioned at the beginning of this chapter, repairing flat-screen TVs has largely become a matter of replacing bad boards or modules, making diagnosis a job of determining which assembly, board, or module is bad.

Try Reset First

Before delving into diagnosis, try performing a reset if there's a reasonable chance a corrupted setting or software could be at fault. There are different types of resets for TVs. The most basic is to unplug the TV for a few minutes and then plug it back in.

The next level of reset restores the TV to its original factory settings. This type of reset involves navigating through the TV menus to locate and perform. The owner's manual will always provide instructions for this.

Some resets can be performed even if the TV won't turn on. These vary considerably by TV and include techniques such as pressing and holding the power button for 10 seconds while the set is unplugged, releasing the button, plugging the set in, and then again pressing the power button for 10 seconds. Use the internet to find a method applicable to your TV.

A service menu reset is more comprehensive than other types of resets. It's intended for professionals and runs the risk of damaging the TV if performed incorrectly. The procedures vary by manufacturer and model and are readily found online.

Software/firmware updating is another way to repair corrupted software problems. Sometimes this can be fully performed through the TV menus of an internet-connected TV. Other times the software will need to be downloaded from the manufacturer's website using your computer and installed via a USB flash drive.

Preliminary Diagnosis

A preliminary problem diagnosis should be performed with the TV back still on. Observe whether the TV shows any signs of getting power, such as an illuminated power LED or an audible click when you plug it in or try to turn it on. If the remote control doesn't do anything, try the power button on the TV itself. If you get a picture, check to see whether the screen is cracked. If you get sound but the screen is completely black, turn out the room lights and see if any light at all seems to be coming from the screen. If not, shine a bright flashlight at the screen a few inches away and look closely to see if there's video activity on the screen. The backlight may be inoperative, but the LCD layer in front of it could be operating properly. If the picture is illuminated but doesn't look right, note whether the entire picture looks bad or only portions of it and the nature of the incorrect appearance.

Secondary Diagnosis

The next step is to take the back off. Once the back is removed, inspect for any visible damage, such as bulging electrolytic capacitors in the power supply board or elsewhere. Look closely for signs of burned or overheated parts. Use caution when working near high-voltage electrolytic capacitors (>50 VDC) in power supply or backlight inverter boards; the caps may be holding a lethal charge even though the TV is unplugged (see "Discharging Electrolytic Capacitors" in Chapter 7).

Appraise the boards and modules in your TV to see if they match the configuration described previously. If you have fewer boards, look to see which functions have been combined. If you have more boards, use online resources to identify them. ShopJimmy.com is a good source of information for TV board identification and replacement. Based on your observations, use Table 9-2 to help identify likely problem suspects and what to do next.

It's usually helpful to do an online search using your model number and symptoms. Some models are notorious for having a particular problem, and there may be an easy fix. Diagnostic blink codes, measurement shortcuts, and reset procedures vary by brand and model, use online resources to find them.

No Signs of Power

If plugging in the TV or trying to turn it on with the power switch on the cabinet does not produce any clicks or power LED activity, the top suspect is the power supply board. Check the fuse on the power supply board, and look for bad capacitors. If the fuse is okay, check the power supply voltages to the main board, especially the standby supply (it's usually labeled). If the standby supply is low or zero, the problem is the power supply board. You'll need to plug in the TV for this step, and there will be potentially lethal voltages on the board. Use an isolation transformer, and keep one hand behind your back when working on a TV that's plugged in. This prevents your opposite hand from completing a high-voltage electrical circuit and passing current through your heart. If the power supply board is okay, the next suspect is the main board.

TV Won't Turn On

If a TV shows any sign of getting power such as a click when you plug it in or an illuminated power LED but won't turn on, the first thing to try is a won't-turn-on reset. You'll need to search online for this type of reset, and it may not exist for your TV. If this doesn't fix the problem, check the power supply voltage to the main board, especially the standby supply (it's usually labeled). If the standby supply is low or zero, the problem is the power supply board. You'll need to plug in the TV for this step, and there will be potentially lethal voltages on the board. Use an isolation

TABLE 9-2 TV Symptoms and Likely Suspects

Symptom	Likely Suspects	What to Do
No signs of power	Power supply board or main board	Check fuse on power supply board. Measure power supply voltages going to main board.
TV won't turn on	Anything	Perform "won't turn on" reset if possible. Measure power supply voltages going to main board.
TV turns off or on by itself	Bad caps in power supply module	Replace bad caps or power supply module.
Blinking power LED	Anything	Find manufacturer codes.
Cracked screen	Screen	Sell modules and stand on eBay, and properly dispose of the rest.
No picture, but backlight working, sound okay	T-con	Replace T-con.
No picture, but backlight working, no sound	T-con or main board	Try replacing T-con.
No picture, no backlight	Backlight power supply, driver, or backlight	Use flashlight to confirm LCD activity.
Entire picture looks bad for all inputs and graphics, no other symptoms	Main board or T-con board	Try replacing T-con.
Only portions of picture look bad	T-con or panel	Try replacing T-con.
Single row of lit or unlit pixels	Panel	Sell modules and stand on eBay, and properly dispose of the rest.
Multiple columns of stuck-on or stuck-off pixels	T-con or panel	Try replacing T-con.
Picture okay, but audio problems	Software corruption or main board	Perform reset; replace main board if reset doesn't fix.
Wi-Fi problems	Software corruption or Wi-Fi module	Perform reset; replace Wi-Fi module if reset doesn't fix.

transformer, and keep one hand behind your back when working on a TV that's plugged in. This prevents your opposite hand from completing a high-voltage electrical circuit and passing current through your heart. If the power supply board is okay, the next suspect is the main board.

TV Turns Off or On by Itself

A TV that turns off or on by itself is almost always caused by bad electrolytic capacitors in the power supply module. Replace the bad caps or power supply module. Discharge high-voltage electrolytic caps in the module before repairing or replacing it.

Blinking Power LED

Some manufacturers build diagnostics into their TVs and use a blinking power LED to tell you what's wrong. These codes vary by model and manufacturer, and you'll need to search online for a key to understanding your code. An example of a blinking-code diagnostic key is in Table 9-3.

TABLE 9-3 TV Blinking Power LED Example Codes

Red LED Blink Count	<Assembly>* Detection Items
2×	<G/B> Main 12-volt overvoltage
3×	<B> Main 5.0-volt failure <B/S> Audio amp. protection <B> Tuner or demodulator I2C no ACK <B> HDMI switch/equalizer I2C communication error
4×	None
5×	<P/T/G/B> Panel ID EEPROM I2C no ACK (also panel power failure is a suspect) <T> T-con IC I2C communication error
6×	<G/P/B> Backlight failure
7×	Overtemperature protection <B> Temperature sensor I2C no ACK <B> V-by-one-lock error between main device and 4KBE device
8×	<B> Software error

*Assembly codes: <G>, power supply board; <B>, main board; <T>, T-con board; <P>, panel module; <S>, speaker.

Cracked Screen

The majority of the cost of a flat-screen TV is the screen, so a cracked screen is usually not worth replacing. Sell the modules and stand on eBay, and properly dispose of the rest.

No Picture but Backlight Working

Detecting whether the backlight is on can be performed in a few ways. In a dark room, it's usually possible to see a small amount of light escaping from the screen itself. Alternately, pinpoints of light can usually be seen through vent holes in the back cover. To be sure, it may be necessary to remove the back cover. If the backlight is working but there's no picture, the T-con board or main board is suspect. If there's sound, this is evidence that the main board is okay.

No Picture, No Backlight

If there's no picture because of the backlight being off, use a dark room and a flashlight to confirm that there's activity on the LCD panel. Press the Menu button on the remote control for this check so that you're sure there's something to display, and shine the flashlight on the screen. If there's LCD activity but no backlight, the backlight power supply, backlight driver, or backlight is suspect. To diagnose these assemblies further, you'll be working near potentially lethal voltages. Use an isolation transformer, and keep one hand behind your back when working near dangerous voltages. This prevents your opposite hand from completing a high-voltage electrical circuit and passing current through your heart.

Entire Picture Looks Bad for All Inputs and Graphics, No Other Symptoms

If you're experiencing video problems that affect the entire picture, check different inputs (HDMI, tuner, USB flash drive with photos) as well as bringing up a menu to display graphics. If the problem only affects some of these, the main board is at fault. If the problem affects them all, the main board or T-con board could be the problem. The T-con board will be less expensive than the main board, so try replacing it first.

Only Portions of the Picture Look Bad
If only portions of the picture (such as the left half of the screen) look bad, the main board is okay. The T-con board or panel is at fault. Try replacing the T-con board.

Single Row of Lit or Unlit Pixels
If only a single horizontal row of pixels is bad, the panel is damaged. Sell the modules and stand on eBay, and properly dispose of the rest.

Multiple Columns of Stuck-On or Stuck-Off Pixels
If multiple vertical columns of pixels are stuck on or off, the main board is okay. The T-con board or panel is at fault. Try replacing the T-con board.

Picture Okay but Audio Problems
If the picture is okay, but you're experiencing audio problems, bad settings or software is a common cause. Perform a reset to restore the settings to their original factory values. If this doesn't work, try updating the software. If this doesn't fix the problem, replace the main board.

Wi-Fi Problems
If you're having Wi-Fi problems, bad settings or software is the likely cause. Perform a reset to restore the settings to their original factory values. If this doesn't work, try updating the software. If this doesn't fix the problem, replace the Wi-Fi board.

Laptop Computers

The most common mishaps with laptops involve cracked screens and damaged keyboards. For some laptops, these are easy and inexpensive to replace. For others, it can be nearly impossible. This is where a site such as iFixit.com comes in handy. The site has model-specific guides for thousands of devices showing exactly how easy or difficult the job is.

Finding Replacement Laptop Parts

When shopping for replacement screens or other laptop parts, search by the model number of the laptop as well as by the many numbers found on the part itself. Some part sellers may use one number for the part, and others may use a different number. Searching different ways will give you the best price and availability. If you don't care whether your screen is glossy or matte, you'll have even more options available.

Be sure that the connectors on replacement screens and keyboards are identical in type and location to those on the original. The same model laptop may have multiple versions using different connectors.

Laptop Cracked Screen Replacement

The following procedure is for a representative laptop and will differ significantly by manufacturer and model. Laptops with large bezels around the screen are usually the easiest on which to replace screens. Refer to Figure 9-4.

1. Turn off the laptop, and unplug the AC adapter. Remove the battery if easily accessible.
2. Remove the plastic bezel by gently popping it off at the top and working your way around the perimeter of the screen.
3. Remove the screws holding the screen to the top cover, as shown in photo b of the figure. Cover the keyboard with a paper towel, and lay the screen onto the keyboard (photo c).
4. The 30-pin connector uses clear tape to keep it from separating. Peel back the tape enough to allow separation, and separate the wiring harness from the screen (photos d and e).
5. Replace the damaged screen with the new one, and reverse the disassembly steps to reassemble the laptop. Be sure that the wiring bundle is pushed into its retainers (photo f).
6. Test the new panel before reinstalling the bezel. If it works properly, reinstall the bezel.

FIGURE 9-4 Laptop cracked screen replacement.

Laptop Keyboard Replacement

The following procedure is for a representative laptop and will differ significantly by manufacturer and model. Refer to Figure 9-5.

1. Turn off the laptop, and unplug the AC adapter. Remove the battery if easily accessible (photo a).
2. Starting at the top-right corner of the keyboard assembly, gently pry up the keyboard while pushing in the adjacent spring-loaded

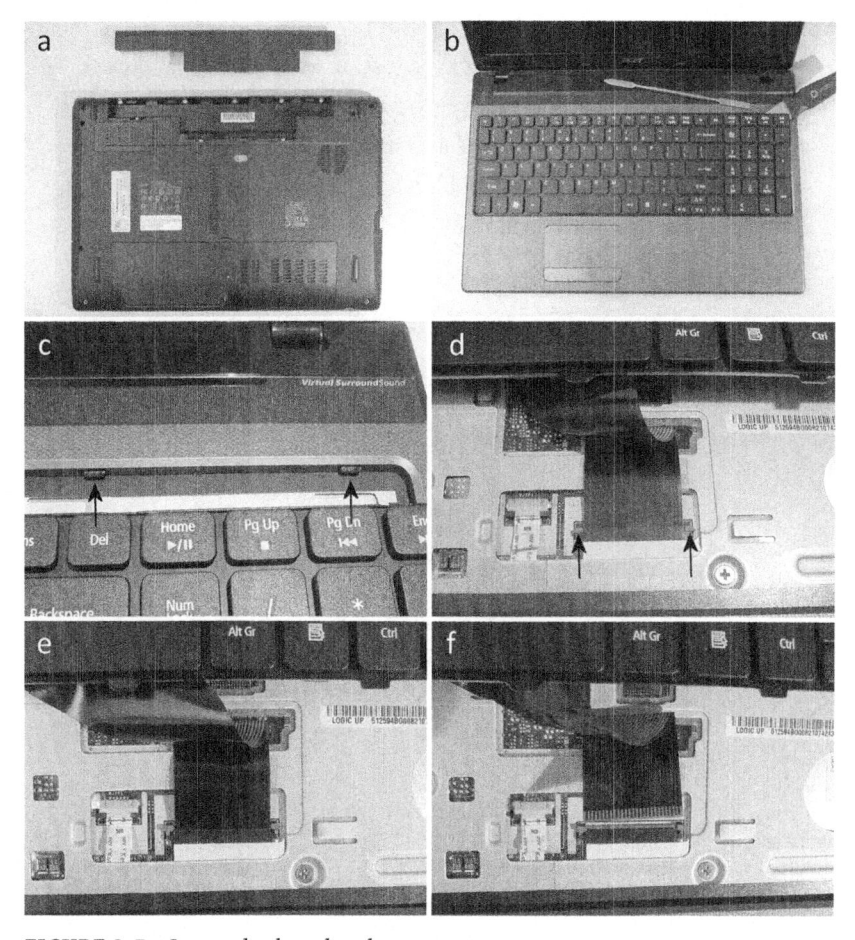

FIGURE 9-5 Laptop keyboard replacement.

retaining tab. Work your way to the left until all tabs are cleared (photos b and c).
3. Lift the top of the keyboard assembly up and forward to expose the keyboard's flexible flat cable connector.
4. Release the flat cable from the connector (photos d–f).
5. Replace the damaged keyboard with the new one, and reverse the disassembly steps to reassemble the laptop.

REPLACING INDIVIDUAL KEYS

If your only keyboard problem is a lost or worn-out key, you're in luck. You can buy and replace single keys without replacing the entire keyboard and save time and money on the repair (Fig. 9-6). Replacing keys will not fix problems where a keystroke is not registered by the computer, though.

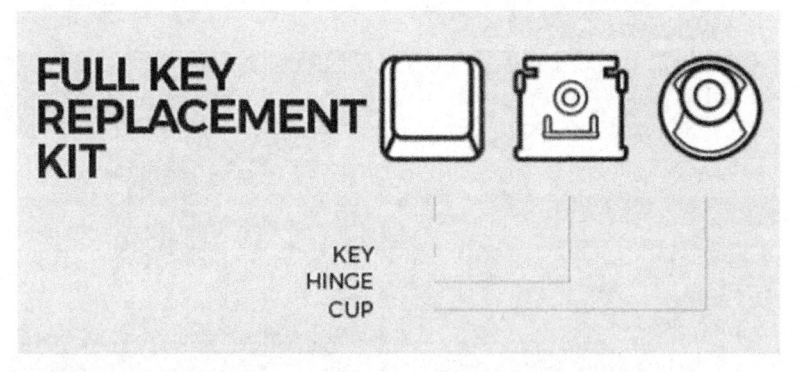

FIGURE 9-6 Laptop key repair kit. *(Courtesy laptopkey.com.)*

Rechargeable Battery Packs

Replace or Repair?

Rechargeable battery packs are found in devices such as cordless drills, work lights, and lawn trimmers. The batteries are good for up to 2,000 cycles, at which point you can either replace the entire device, replace the battery cartridge (if it has one), or repair/rebuild the battery pack within the device or cartridge using replacement cells.

To figure the cost of repairing or rebuilding a battery pack, you'll need to identify the number and type of cells.

Sometimes just a single cell or two have gone bad, and it's tempting to replace just the bad ones. Bad cells can be located by measuring the DC voltage of each cell and looking for the ones that have zero or nearly zero volts across them. In practice, once one cell fails, others are usually close behind. Replacing a bad cell may buy another few months of battery pack use, but don't expect much more than that. The same advice applies to restoration techniques such as applying a large current pulse to a bad NiCd cell to vaporize internal dendrites that have formed.

Identifying Cell Types

Whether replacing one or all of the cells in a rechargeable device, you'll need to know their chemistry, capacity, and physical dimensions.

Chemistry

There are two common rechargeable battery chemistries suitable for cell replacement:

- Nickel-cadmium (NiCd)
- Nickel–metal hydride (NiMH)

A third common rechargeable battery chemistry, lithium-ion (Li-ion), is not a good candidate for cell replacement. Li-ion replacement cells are not readily available, their cell voltages differ by exact chemistry, and the cost of rebuilding a battery pack usually exceeds that of a new one.

Battery chemistry is usually shown on the product as well as stamped on individual cells. NiCd and NiMH batteries nominally output 1.2 volts per cell. This allows you to calculate how many cells are in a device without taking it apart. Do this by dividing the voltage of the device by the cell voltage. For example, an 18-volt NiCd cordless drill would use 18 volts/1.2 volts = 15 cells.

There are important differences in how each of the battery chemistries is charged to maximize performance and lifespan and avoid overheating hazards. This means that a charger designed for one chemistry may not be suitable for another. For this reason, it is best to replace batteries with the original chemistry.

Capacity

Capacity, measured in milliamp-hours (mAh), tells you how many milliamperes of current a battery can supply for 1 hour. For example, a 1,000-mAh battery can supply 1,000 mA for 1 hour. This same battery could instead provide 2,000 mA for ½ hour or 500 mA for 2 hours. The capacity rating tells you the available current multiplied by the duration of use.

The capacity is generally stamped on the cells. If you are replacing a single or a few cells, use replacements having very close to the same milliamp-hours as the originals. If you're replacing all the cells, you can increase the milliamp-hour rating to provide a longer run time for your device. Most chargers are forgiving of modest increases in battery capacity.

Dimensions

Dimensions for rechargeable batteries are often provided in inches or millimeters, but standard battery designations are also used to specify size. In addition to AAA, AA, and C, there are sizes such as ⅔AA and the very popular sub C (Fig. 9-7). Sub C batteries are 1.7 inches long and 0.9 inch in diameter. You'll usually need to replace cells with ones having the same size as the originals in order to fit in the device.

FIGURE 9-7 Sub C batteries with solder tabs.

Rebuilding a Battery Pack

Follow these steps and refer to Figure 9-8 to rebuild a battery pack:

1. Disassemble the battery cartridge or device to reveal the batteries (photo b of the figure). Determine their chemistry, capacity, and physical dimensions. Order cells as needed.
2. Arrange the cells in the same way as the originals, aligning the positive solder tab of one cell with the negative solder tab of the next cell. Use tape or hot glue to maintain alignment (photo c).
3. Solder the mating solder tabs together, and solder the battery connection wires to the outermost solder tabs (photos d and e).
4. Reassemble the cartridge or device by reversing the disassembly steps.

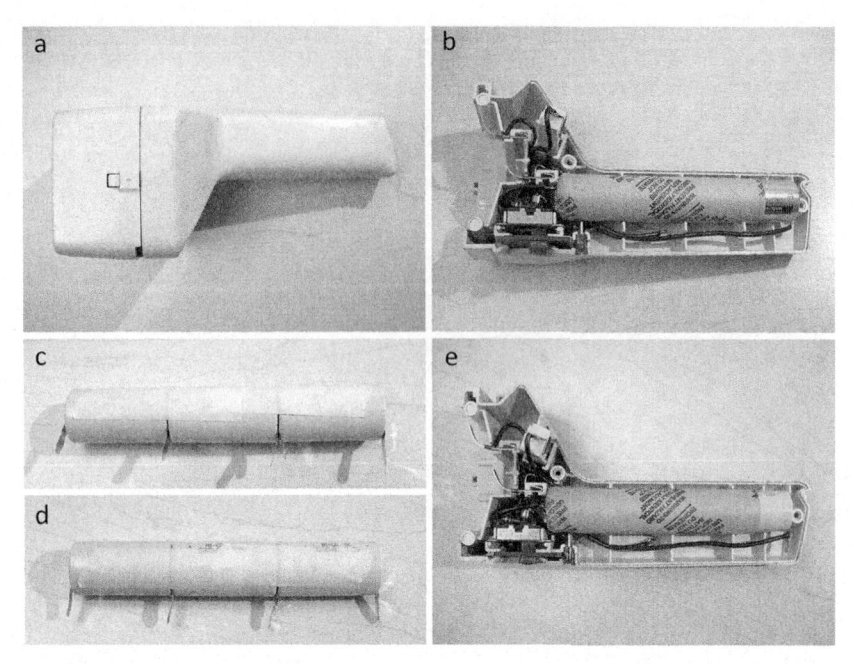

FIGURE 9-8 Rebuilding a battery pack.

PROPERLY DISPOSE OF USED RECHARGEABLE BATTERIES

NiCd is considered hazardous waste and must be taken to an e-recycler. NiMH is considered semitoxic and Li-ion is considered mildly toxic; both are best e-recycled.

If enough batteries contact each other, they can cause a fire, even if you think they don't have any charge remaining in them. To prevent problems, put some tape over the positive ends to prevent current from flowing (Fig. 9-9).

FIGURE 9-9 NiCd cells with taped ends.

Audio Receivers and Power Amps

The two most common problems with audio receivers and power amps are bad electrical contacts or blown output amplifier circuitry.

Electrical Contact Problems

Bad electrical contacts can be in selector switches, input and output jacks, and controls such as volume and balance. Contact problems are often intermittent and are common in vintage audio gear. Electrical contacts oxidize over time from exposure to the air, resulting in intermittent or bad connections. In mild cases, this produces scratchy volume and tone controls. In more extreme cases, equipment just does not work.

Electrical contact cleaner sprays are the closest thing to a repair kit in a can for contact problems. I use CAIG DeoxIT D5 for switches and connectors and DeoxIT Fader F5 for pots (potentiometers). I've had pieces of equipment that were completely nonfunctional until contact spray treatment.

The trick with contact sprays is to get the chemical into the contacts without getting it everywhere else. The CAIG FLEX-TIP helps with that. Remove the cover of your receiver or amp to get access to all the parts that you want to treat. Look for vent holes in can covers of potentiometers and side openings of switches to spray into. After spraying, work the switch or pot back and forth over the entire range many times to distribute the chemical and scrub the contacts. Clean up the excess and let dry.

Another possible culprit for intermittent outputs in audio receivers and amplifiers is relays for speaker protection and speaker channel switching. Check the solder connections of relays, and use an oscilloscope to look at the signals on both sides of closed relay switches to make sure that they look identical.

Blown Output Amplifier Circuitry

The output stage of amplifiers can fail when speaker wires are accidentally shorted together, low-impedance loads are connected, or when the unit is subjected to extended periods of excessive output power. Output stages can use discrete transistors or integrated circuits. High-power designs favor discrete transistors because there are practical limits to managing heat with an integrated circuit package. Look at what's mounted on the big heat sink to indicate whether the amp uses an integrated or discrete output stage.

Amplifier Class

Audio amplifiers are most commonly Class AB or Class D. Either class can use discrete transistors or integrated circuits for the output stage. Class AB is a classic linear amplifier that offers excellent performance at reasonable efficiency. Class D is a relative newcomer to high-end, full-bandwidth audio. It's a switching amplifier that offers superior efficiency, which translates to less heat. The disadvantage is the large amount of radio-frequency energy internally generated as a by-product of switching. This must be controlled by shielding to prevent interference with radios and wireless devices.

Speaker Protection Relays

Many receivers and amplifiers use relays to disconnect speakers from the amplifier outputs until a few seconds after turn-on, at turn-off, and if a fault condition such as DC voltage at a speaker output is detected. You can often hear the relays click shortly after device turn-on.

When troubleshooting amplifier outputs, be aware that a fault-activated or damaged relay may be decoupling the amplifier from the speaker outputs.

Diagnosing Amplifier Output Stages with Integrated Circuits

If you are working on output stages using integrated circuit (IC) power amps, measure the DC voltages on each of the pins of the heat-sink-mounted ICs, comparing them against the expected value from an IC datasheet. If no datasheet is available, compare one channel against another, looking for significant differences. If no signs of component damage are seen in nearby circuitry, voltages that are different than expected are evidence that the IC has failed.

Use caution when measuring voltages on IC amplifier pins; many of them are lethally high. It's best to use clip leads to make voltmeter connections with the power off—it's too easy to make a slip shorting two pins together and damage something with the power on. Heat sinks in audio receivers and amps are generally grounded, but measure their voltage with a DC voltmeter to be sure before measuring anything else. Follow safe high-voltage working practices such as working with one hand behind your back.

Follow the guidance in the "Heat-Sinked Devices" section of Chapter 5 when replacing IC amplifiers.

Diagnosing Amplifier Output Stages with Discrete Transistors
If you are working on output stages using discrete transistors, check all the heat-sink-mounted transistors for faults. Start by checking each transistor for shorts between any of its pins. Compare results from one channel to another as a sanity check. If only one channel is bad and left and right channel transistor counterparts measure the same, the problem may be elsewhere.

If in-circuit testing for shorts is inconclusive, turn on the power and use the techniques in the "Transistor Circuitry Troubleshooting" section of Chapter 8. Use caution when measuring voltages on output transistors; many of them are lethally high. It's best to use clip leads to make voltmeter connections with the power off—it's too easy to make a slip shorting two pins together and damage something with the power on.

Heat sinks in audio receivers and amps are generally grounded, but measure their voltage with a DC voltmeter to be sure before measuring anything else. Follow safe high-voltage working practices such as working with one hand behind your back.

If an output transistor is blown, the transistor that drives it may be blown too. If you don't have a circuit diagram, follow the wiring from the base or gate of the output transistor until you find it. Also check the fractional-ohm power resistors between the output transistor and the speaker terminal.

Follow the guidance in the "Heat-Sinked Devices" section of Chapter 5 when replacing heat-sinked transistors.

Bias Adjustment Following Transistor Replacement
If you've replaced a transistor in a Class AB output stage and the amplifier is working again, there's still one more thing to do: adjust the bias current. The bias current is the current that flows through the output transistors even with no input signal. Too little bias current produces distortion; too much wastes power and reduces reliability. The setting that provided the proper bias for the original transistor may not provide the same bias current for the replacement transistor.

The following procedure is for a representative bipolar transistor (NPN or PNP) amplifier and will differ by manufacturer and model. You'll want to find the service manual for your amplifier to perform the adjustment correctly. If the service manual is not available and the bias potentiometer is clearly labeled on the PC board, you can use the undamaged channel as a current reference for checking the repaired channel.

1. With 8-ohm resistive loads and no signal in, set bias control(s) to midpoint.
2. Connect a DC millivolt meter across any of the output emitter resistors for the channel under alignment.
3. Turn amplifier on, and allow a 3- to 5-minute settling period.
4. Adjust the bias control to obtain either a + or − 16-mV indication on the millivolt meter. This corresponds to 20 mA of bias current.
5. To check for proper bias setting and tracking operation, remove the millivolt meter, and apply an input signal to reach 66 watts into 8 ohms for 10 minutes with the cover on.
6. Remove the input signal, and quickly connect the millivolt meter as in step 2. The meter should indicate 50–60 mV, decreasing as the amplifier cools down.
7. Compliance with step 6 indicates proper bias alignment. If not, unplug amplifier, allow it to cool down, and repeat the procedure.

If you have a function generator and oscilloscope, another method that can be used to adjust bias is to provide a sine-wave input to the amp or receiver and monitor the signal at the speaker output with the scope. Increase the bias to the point where the switching distortion just disappears at the zero crossings of the sine wave.

WHY IS BIAS ADJUSTMENT NEEDED?

Figure 9-10 shows a Class AB complementary amplifier output stage. *Complementary* means that matching NPN and PNP transistors are used to source and sink current at the speaker output. The output stage is driven by preamp and bias circuitry.

Music produces both positive and negative output swings. For positive swings, transistor Q3 provides the positive current to the speaker output. For negative swings, transistor Q4 provides the negative current to the speaker output. As the output approaches zero, there's a handoff between the two transistors. The bias current determines the overlap in the operation of the two transistors to provide a smooth handoff.

If the bias current is too low, one transistor turns off before the other turns on, producing audible distortion at the switching point. If the bias current is too high, the output transistors will run hotter than necessary. This wastes power and reduces their reliability. Adjusting the bias current is a compromise between these two extremes.

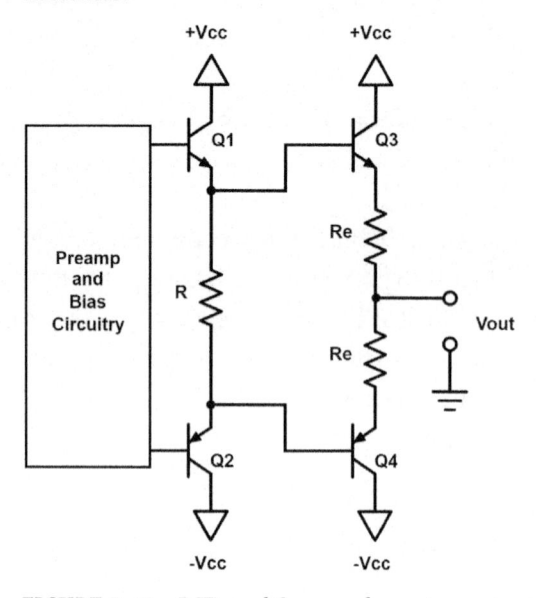

FIGURE 9-10 BJT amplifier complementary output stage.

Loudspeakers

The two most common types of loudspeaker problems are drivers damaged by too much power (usually tweeters) and deteriorated foam woofer surrounds. Both types of repairs will be covered here.

Driver Damage

Most loudspeakers today are either two-way (woofer for the bass and tweeter for the treble) or three-way (adds a midrange driver). Whether a speaker is considered two or three-way depends on how many frequency ranges are used, not the number of drivers (Fig. 9-11). Tweeters blow most often—and for good reason. When the volume of music is turned up to the point of amplifier distortion, most of that distortion is in the form of high-frequency energy, which rapidly overloads tweeters.

FIGURE 9-11 Three-way speaker anatomy.

In the case of woofers, turning up the bass too loud can either damage the woofer suspension system by mechanical overexcursion or scorch the voice coil with more power than it can safely dissipate.

Blown tweeters are easily detected by a listening test, or you can use an ohmmeter to see if they're open (infinite resistance). Damaged woofers will produce distorted sound when playing music, and you can hear the voice coil rub if you push in on the center of the cone with your hand.

Driver Replacement

In a few cases, such as profession sound reinforcement speakers, you can buy replacement tweeter voice coils and repair blown tweeters. Replacement voice coils are also available for some high-end car audio subwoofers, but most of the time you'll have to replace the entire driver.

It's always best to replace drivers with the originals, but they can be prohibitively expensive or impossible to find. Madisound.com maintains a list of proven driver substitutes for high-end speakers. The last resort is finding your own substitute. Madisound and Parts Express are two good sources for drivers.

When substituting drivers, both mechanical and performance attributes are important. Ideally, a substitute will have a similar appearance, fit the same mounting hole, and have the same screw-hole locations. Matching performance is a bigger challenge. The impedance of the replacement should match that of the original. The impedance value is almost always stamped on the back of drivers; 4 and 8 ohms are most common. Choose a replacement of the same format. If the original is a 1-inch-diameter dome, pick a 1-inch-diameter dome replacement. Try to find replacements that have a power rating equal to or higher than that of the original. Tweeter power can be specified in one of two ways, actual power or frequency-rated power. An actual power rating might only be a few watts. Frequency-rated power ratings consider the fact that tweeters are used with crossover filters that block low-frequency energy, so a tweeter may only need to handle 5 percent of the total. Thus, the same tweeter could have a 5- or a 100-watt rating. Woofers and midranges only use the actual power rating.

Always replace drivers in pairs. It's important that the left and right speakers sound identical to each other, even if the substitutes don't sound as good as the originals. Speakers have a polarity to them, so connect the replacements the same as the originals. Positive terminals may be marked with a plus (+) sign, a dot, or be colored red.

Listening Test and Fine-Tuning

If you've substituted drivers, you may have altered how your speakers sound. Trust your ears to tell you how you've done by listing to your favorite music on them. There are a few adjustments that can be made that may improve the result. Reversing the polarity of the tweeter affects the frequency overlap region between the tweeter and woofer (or midrange in a three-way design). It can increase or decrease the output response in the frequency overlap region. The polarity that provides the perceived highest output response is usually best.

If you've replaced a tweeter and it seems too loud, you can reduce its volume by using a resistive pad. A pad requires two resistors and reduces the volume of the tweeter without changing the impedance that the crossover network sees. Using a single resistor in series with the tweeter would reduce the tweeter level, but it would also lower the crossover frequency. Figure 9-12 shows 3-decibel (dB) pads for 4- and 8-ohm tweeters. This design will reduce the power to the tweeter to the proper amount for most situations where a substitute tweeter is too loud. The power ratings shown are for a 100-watt system. You can adjust the resistor power ratings accordingly for systems having less or more rated power handling. For example, a 50-watt system would let you cut the resistor power ratings in half.

3 dB tweeter pads 100 watt system

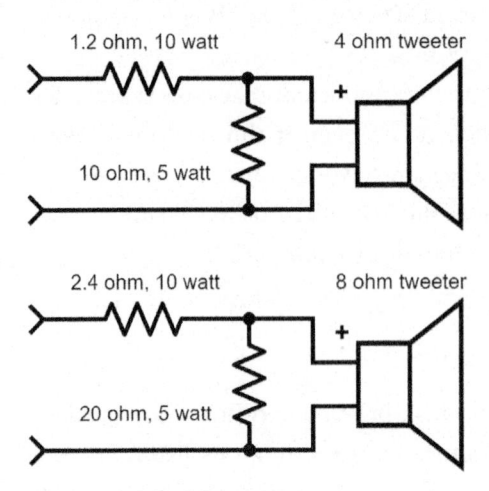

1.2 ohm, 10 watt — 4 ohm tweeter
10 ohm, 5 watt

2.4 ohm, 10 watt — 8 ohm tweeter
20 ohm, 5 watt

FIGURE 9-12   3-dB pads for 4- and 8-ohm tweeters.

Additional Considerations for Woofer Substitution

Woofers have three additional parameters that affect bass performance. They're collectively known as the *Thiele/Small (T/S) parameters* and are named fs, Qts, and Vas. Even if we knew what the original woofer T/S parameters were, it's not important to match them. What's important is that the replacement woofer is compatible with the speaker box.

Sealed boxes (i.e., no vent or port) are forgiving of a wide range of T/S parameters. The best simple advice for them is to avoid choosing a replacement woofer that has a Qts value larger than 0.7. Woofers with high Qts values have weak magnet structures and run the risk of producing boomy, resonant bass.

Ported speaker boxes are more complex in design and are much more sensitive to T/S parameter changes than sealed boxes. If you want to do a good job substituting a woofer in a ported box, you'll need to use loudspeaker box design software to check and compare the performance of candidate woofers in your particular box. You'll need to enter the cabinet and port dimensions of your box as well as the T/S parameters of the woofers you're considering as substitutes and then evaluate the results. The Subwoofer Design Toolbox by MFR Engineering is an easy-to-use, affordable tool for this and works for any woofer, not just subwoofers. Using software also benefits sealed-box woofer substitution projects.

Woofer Surround Repair

A woofer surround provides a flexible interface between the woofer cone and frame. It keeps the cone centered and provides an airtight seal. Figures 9-13 and 9-14 show the anatomy of a woofer, including the surround.

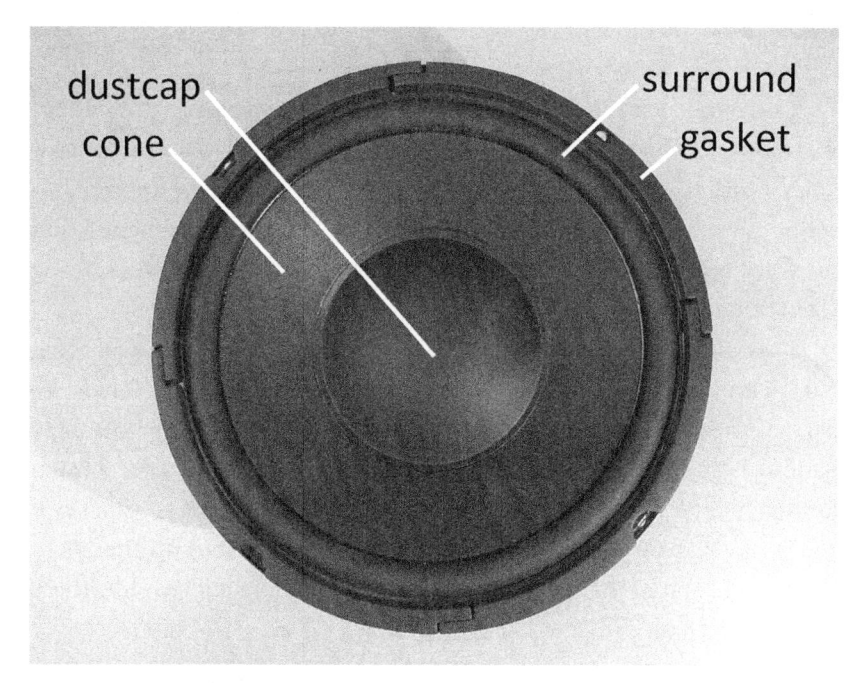

FIGURE 9-13 Woofer anatomy, front view.

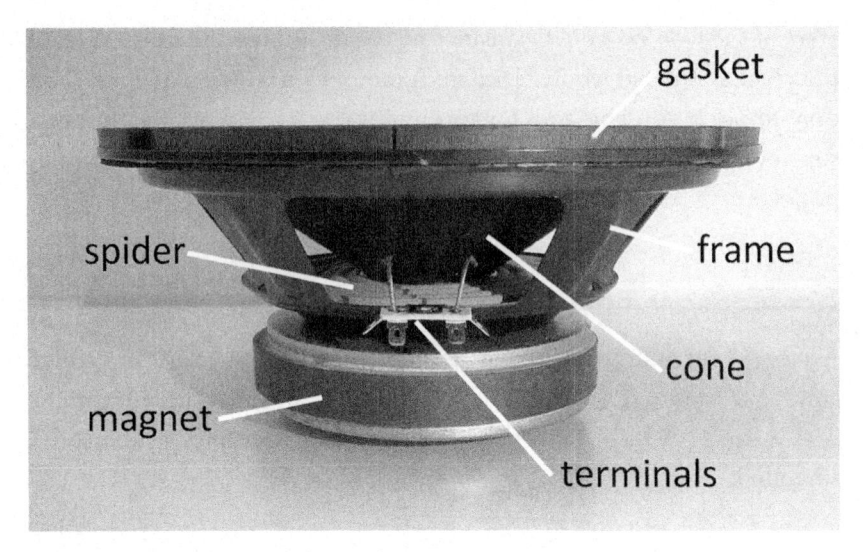

FIGURE 9-14   Woofer anatomy, side view.

Surrounds can be made of rubber, foam, or other materials. Surrounds made of foam eventually deteriorate and develop tears or holes. Once this happens, the woofer is at high risk of mechanical damage because the damaged foam can't keep the cone properly aligned, and the voice coil assembly at the base of the cone can begin to rub, causing voice coil damage.

Fortunately, it's possible to replace the foam surround on any woofer, but it's important to check beforehand whether the woofer is otherwise okay. If the foam has just started to develop minor tears or holes, the outlook is good. Play music through the speaker in its cabinet and listen for distortion or rubbing. Compare it against its companion speaker if you're having trouble deciding if there's a problem. If the surround is so badly deteriorated that you can wiggle the cone from side to side, there may be voice coil damage.

To Shim or Not to Shim

The basic idea behind replacing a woofer surround is simple: remove the old surround, and glue in a fresh one. The tricky part is making sure that the new surround is glued in so that the cone is perfectly centered. If it isn't, the voice coil will rub, and you'll have to redo the whole procedure. There are two techniques for replacing surrounds, shims, or shimless (Table 9-4).

TABLE 9-4 Advantages and Disadvantages of Shimming

Technique	Advantage	Disadvantage
Shims	Very low risk of having to redo	Must replace dust cap
Shimless	Fewer steps can be used with all woofers.	Requires more finesse, with a higher risk of having to redo.

Shims and voice coil rub can be understood by looking at Figure 9-15. Diagram a shows the parts of a woofer, including the voice coil and pole piece. These are the parts that can rub together if the cone isn't

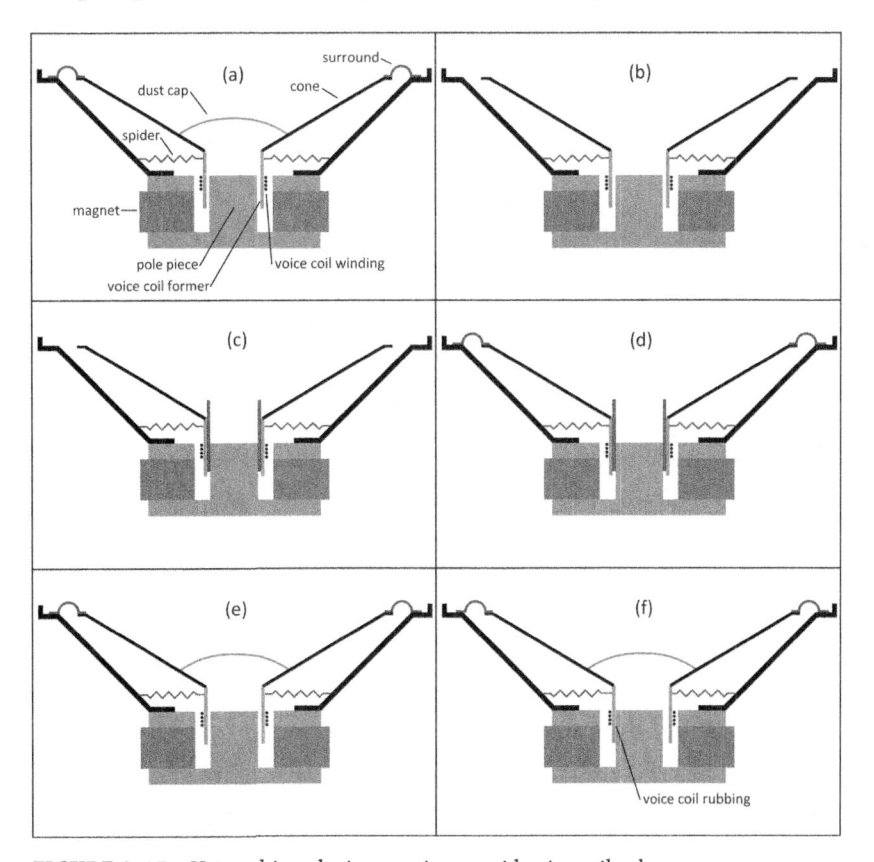

FIGURE 9-15 Using shims during repair to avoid voice coil rub.

centered, as shown in diagram f. Note that the normal space between the two is greatly exaggerated in the figure. In practice, it may not be much more than the thickness of a few sheets of paper.

Figure 9-15, diagrams b–e show the standard procedure for replacing a surround using shims. Diagram b shows the old surround and dust cap removed. Diagram c shows shims evenly inserted into the space between the voice coil and the pole piece. Diagram d shows the new surround glued on. Diagram e shows the shims removed and a replacement dust cap glued on. If you buy a surround repair kit (Fig. 9-16), it will come with everything you need, including glue, shims, dust caps, and detailed instructions for doing a repair with shims.

FIGURE 9-16 Woofer surround repair kit. *(Courtesy of Parts Express.)*

Sometimes woofers don't have a removable dust cap, for example, the woofers shown in the three-way speaker anatomy photo. Other times the dust cap has a prestigious logo on it you don't want to lose. In such cases, going shimless is the best option. The procedure will be explained later.

Surround Size and Edge Type

Regardless of whether you use shims or not, you'll need to buy the right size and type of replacement surrounds. If you're repairing a popular woofer, the homework has already been done for you, and you can just order Bose 901 or JBL 125A surrounds. If not, you'll need to get out your ruler.

Figure 9-17 shows the two surround edge types. The surrounds need to match the cone edge type.

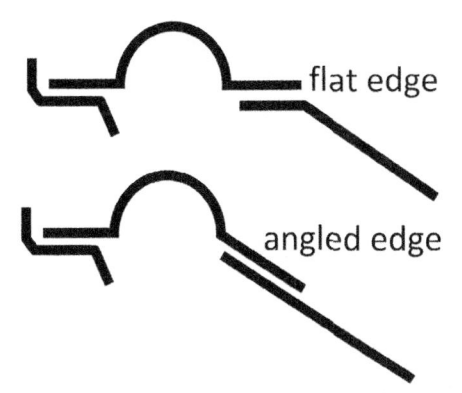

FIGURE 9-17 Surround edge types.

The surround size needs to match, too. Unfortunately, there isn't much standardization here, and what one company calls a 10-inch woofer may be a different size from what another company calls a 10-inch woofer. You'll need to measure the four dimensions shown in Figure 9-18 as A, B, C, and D. Dimensions B and C are the most important and need to be closely matched between woofer and surround. Dimensions A and D only affect the widths of the glueable surfaces and don't need to be perfect matches.

If you're not able to find a replacement surround with acceptable dimensions and edge type, you have a few options. One option is to use a surround with the incorrect edge type. Another option is to slightly reduce the diameter of a replacement surround by cutting out a small section and gluing the ends together, as shown in Figure 9-19. Overlap the two ends by about ½ inch. You'll want to buy an oversized surround with the same *roll width* as the original. This means that the difference between dimensions B and C should be the same as the original.

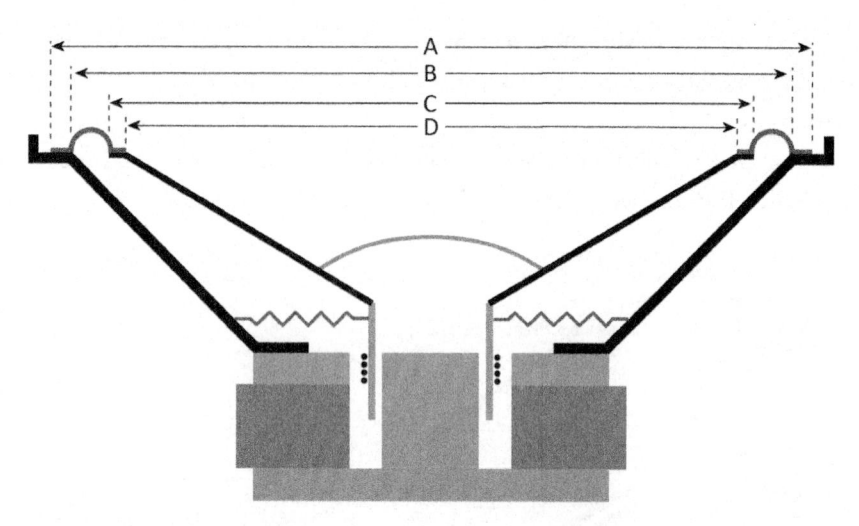

FIGURE 9-18 Surround dimensions.

FIGURE 9-19 Resized surround.

Shimless Surround Repair

Follow these steps to perform a shimless surround repair. Refer to Figure 9-20. The woofer should first be checked to make sure that the centered cone can be moved up and down without rubbing.

1. Remove the woofer from the speaker enclosure, and place it on your workbench (photo a).
2. If there's a top gasket (as shown in the woofer anatomy photos), carefully remove it. It will need to be reused.
3. Remove the old surround from the cone and frame (photo b). It's not necessary to remove the adhesive residue unless it's tacky. Any solvents should be used in a well-ventilated area and tested to make sure that they don't damage the woofer cone.
4. Glue the new surround to the top or bottom of the cone, the same as the original. Allow the glue to dry thoroughly.
5. Position the cone with attached surround as far toward the 12 o'clock position as it will easily go; then make a mark on the frame using a fine-tip paint pen or Sharpie at the edge of the surround at that spot. Repeat for 6 o'clock, 3 o'clock, and 9 o'clock. Repeat for the in-between positions (1:30, 4:30, 7:30, and 10:30). You now have eight marks establishing cone rub limits (photo c).
6. Glue the outer edge of the surround to the frame, being careful to center the outer surround edge between the eight marks made in step 5. If you can tell that the voice coil is rubbing (listen while moving the cone up and down), slightly shift the outer edge of the surround in the direction that makes the rubbing disappear. Allow the glue to dry thoroughly.
7. Test the woofer using music with the bass turned up or bass tones from a signal generator. The test signal should be loud enough to cause the woofer cone to move over its useful excursion range. Listen for any signs of rubbing. If it sounds good, glue the top gasket back on (photo d) and reinstall it in the speaker enclosure. If you hear rubbing, you'll need to repeat the repair.

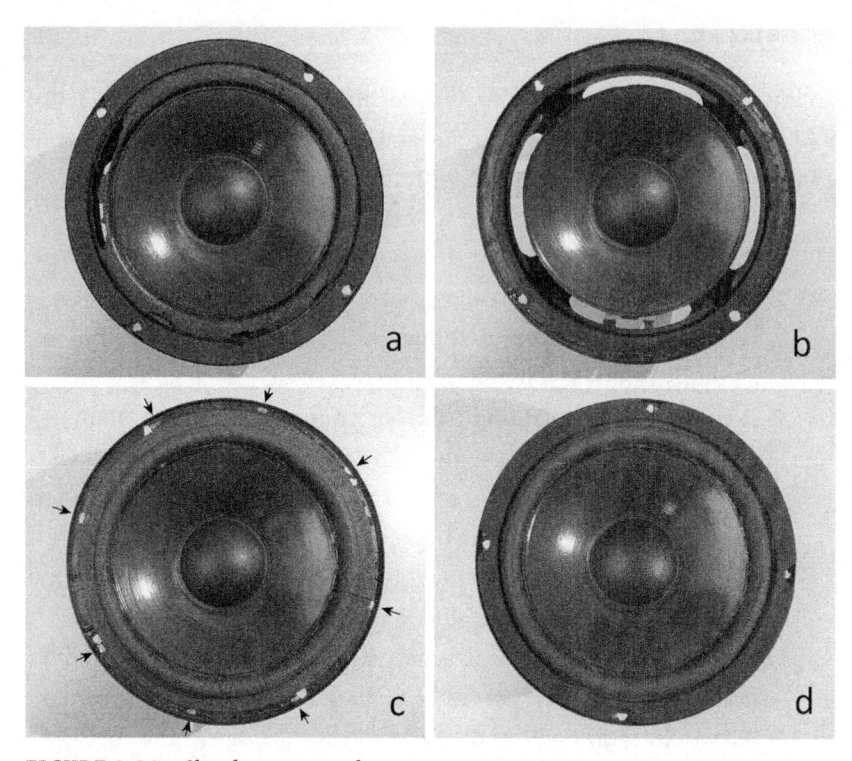

FIGURE 9-20   Shimless surround repair.

Remote Controls

Remote control problems are usually either user error, battery problems, or keypad problems.

User Error

First, make sure that the remote is in the right device mode. Sometimes an incorrect device button gets pressed accidentally, and the user doesn't realize that the remote is communicating with the Blu-Ray player instead of the TV. This is made more confusing by the volume punch-through mode of many multidevice remotes that operate the TV or audio receiver volume controls even when the remote is in another device mode.

Product-Specific Repairs | 229

Battery Problems

If the remote seems dead, remove the batteries and inspect the contacts in the remote. If there are any signs of corrosion, use the eraser on the top of a pencil to scrub them. If there's battery leakage, clean the contacts with damp cotton swabs and a dental pick. Put in fresh batteries, making sure that they're installed correctly. Usually, the springs go to the negative battery terminals, but not always. Trust the plus and minus signs not whether there's a spring.

If this doesn't fix the problem, use the camera test for remote controls described in the sidebar to determine whether the remote is sending out infrared commands or not.

CAMERA TEST FOR REMOTE CONTROLS

Infrared (IR) remote controls send out IR light pulses to control their companion devices. The human eye can't see IR light, but many digital cameras found in phones, tablets, and laptops can (Fig. 9-21).

FIGURE 9-21 Camera test for remote controls.

In a dark setting, point the remote control toward the camera, and press the volume-up button. (Volume-up keeps sending pulses as long as you press the button.) If the remote is working and your camera is sensitive to IR light, you'll see a glow from the IR LED in the remote. If you don't see a glow, try another camera device or another remote control until you're sure that you have a suitable camera for the test.

Keypad Problems

Keypad problems require taking the remote control apart. Hidden tabs are commonly used to hold the two halves of the case together. Many remotes also use screws, so if a remote seems like it's not coming apart at a certain spot, there's probably a missed screw. Prying apart remotes requires finesse to avoid damage. You can buy a pry tool set to make the job easier. If you don't have a pry tool set, credit cards and guitar picks come in handy.

Once the remote is apart, separate the rubber keypad membrane from the PC board beneath it. Gently clean both the underside of the rubber keypad and the mating PC board contacts using isopropyl alcohol and cotton swabs (Fig 9-22).

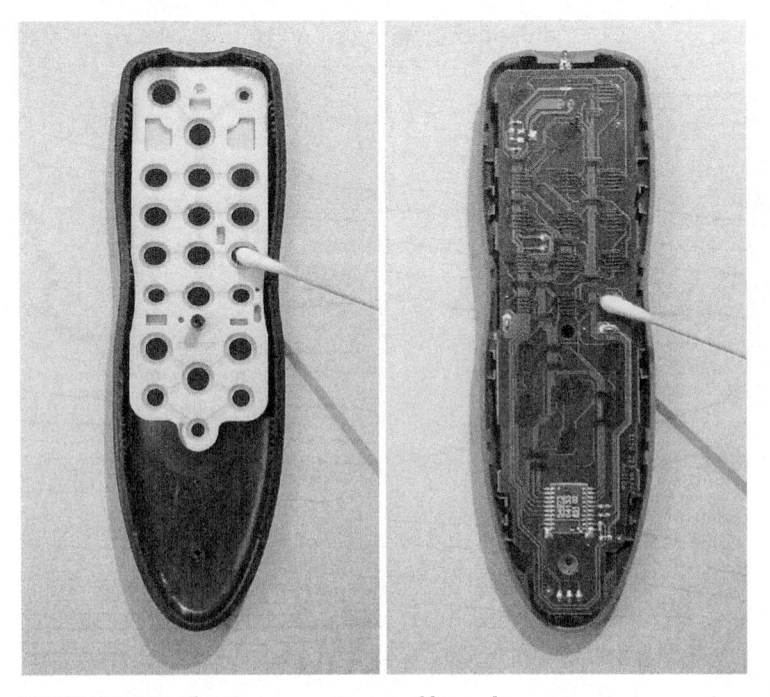

FIGURE 9-22 Cleaning a remote control keypad.

Replacement

If you're not able to repair your remote, your options are to buy an original replacement, a related-model replacement, a custom replacement, or a universal remote. Custom replacement remotes come preprogrammed to provide the same functions as the original but have a different look (Fig. 9-23). Think of them as universal remotes that have all the right buttons and that you don't need to program. Universal remotes are inexpensive, but you'll need to program them, and they may not have all the functions of the original.

FIGURE 9-23 Redi Remote custom replacement remote. *(Courtesy of remotes.com.)*

Part Suppliers

Electronic Components

Digi-Key Electronics, https://www.digikey.com
Mouser Electronics, https://www.mouser.com

TV Repair Parts

ShopJimmy, https://www.shopjimmy.com

Loudspeaker Parts and Supplies

Madisound, https://www.madisoundspeakerstore.com
Parts Express, https://www.parts-express.com

Electronics Repair Tools and Parts

iFixit, https://www.ifixit.com

References

Chapter 5: Soldering and Desoldering

Bixenman, M., et al. "Why Clean a No-Clean Flux." *Proceedings of the International Conference on Soldering and Reliability (ICSR)*, 2016; available at www.kester.com/Portals/0/Documents/Knowledge%20 Base/Publications/International%20Conference%20on%20Soldering %20and%20Reliability%20ICSR%202016_paper_Why%20Clean%20 A%20No-Clean%20Flux.pdf.

"Model WLC100 Soldering Station Manual." Earl E. Bakken Medical Devices Center, University of Minnesota; available at www.mdc.umn .edu/facility/files/electrical/ElecManuals/WellerSolderingStation Manual.pdf.

Joel, E. B. "How to Solder: Through-Hole Soldering." SparkFun, learn.sparkfun.com/tutorials/how-to-solder-through-hole-soldering/ all.

Chapter 6: Using Test Equipment

"FLUKE 114, 115, and 117: True-Rms Multimeters User's Manual." 2006 (revised January 2, 2007). Fluke-Direct.com, www.fluke-direct .com/pdfs/cache/www.fluke-direct.com/114-efsp/manual/114-efsp -manual.pdf.

Tektronix. "TBS1000B and TBS1000B-EDU Series Oscilloscopes User's Manual." 2007. Tektronix, www.tek.com/oscilloscope/tbs1000b-edu-digital-storage-oscilloscope-manual/tbs1000b-and-tbs1000b-edu-series-oscil.

Chapter 7: Part Identification and Substitution

EETech Media. "The Resistor Guide: Your Guide to the World of Resistors." 2019. Available at www.resistorguide.com/.

M, Laura. "Semiconductor Counterfeiting Is a Global Problem." July 31, 2017. SiliconExpert Blog, www.siliconexpert.com/blog/semiconductor-counterfeiting-global-problem/.

Rubycon Corporation. "Technical Notes for Electrolytic Capacitor." Available at rubycon.co.jp/en/products/alumi/pdf/Performances.pdf.

Sphere Research Corporation. "The SMD Codebook." Available at www.sphere.bc.ca/download/smd-codebook.pdf.

Chapter 8: General Repairs

Agans, D. J. *Debugging: The 9 Indispensable Rules for Finding Even the Most Elusive Software and Hardware Problems.* New York: AMACOM Books, 2006.

Geier, M. J. *How to Diagnose and Fix Everything Electronic,* 2nd ed. New York: McGraw-Hill Professional, 2015.

Chapter 9: Product-Specific Repairs

"BU-705: How to Recycle Batteries." September 13, 2019. Cadex Electronics, batteryuniversity.com/learn/article/recycling_batteries.

Index

AC adapters, 144–149
    regulated and unregulated, 144–145
    repair, 146–147
    replacing DC power connectors, 147–148
    substitution, 149
    testing, 145–146
AC voltage, 22–23
AGC (see fuses)
AGX (see fuses)
air duster, 26–27
alternating current
    (*See* AC voltage)
amplifier circuit, 186–187
amplifier Class AB, D, 212–213
analog meters, 141–143
    replacement, 143
    stuck needle fixes, 143
    substitution, 143
    testing, 141
antistatic mat, 53–54
antistatic wrist strap, 53–54
Apollo 11, 191
audio receivers and power amps, 211–216
    bias adjustment, 214–216

blown output amplifier circuitry, 212–216
    diagnosing IC output stages, 213
    diagnosing transistor output stages, 214
    electrical contact problems, 211–212
axial leads, electrolytic capacitors, 106–107

back flip connectors, 161–162
batteries:
    checking alkaline, 89
    in series and parallel, 20–21
battery terminals, cleaning, 153
BGA (*See* integrated circuits, packages)
bias adjustment, in audio receivers and power amps, 214–216
bipolar transistor amplifiers, troubleshooting, 182–183
bipolar transistors (NPN, PNP), 121
    testing, 121–122
block diagrams, 16–17
board connectors, 160–163

Index

bulging electrolytic capacitors, 164
burned components, 165
bypassing, 112

cable ties (*See* wire ties)
calipers, digital, 55–56
camera test for IR remotes, 229
capacitance meter (*See* ESR meter)
capacitors:
   ceramic, 113–115
   electrolytic, 5–6
   electrolytic, 106–111
   film, 113
   identification, 106
   in series and parallel, 22
   microwave oven, 110
   tantalum, 111–112
   uses, 11
caps (*See* capacitors)
CCFL (cold-cathode fluorescent lamp) backlights, 192–193
ceramic capacitors, 113–115
   class 1, class 2, 113–114
   marking code, 114
   replacement, 115
   temperature coefficient, 115
childproofing work area, 8
chokes (*See* inductors)
circuit diagrams (*See* schematic diagrams)
Class AB amplifier, 212–213
   complementary output stage, 216
Class D amplifier, 212–213
clip leads, 35–36
coefficient of thermal expansion, 168
coils (*See* inductors)

cold solder joints, 70
complementary output stage, 216
component symbols, 18
connectors, 160–163
   back flip, 161–162
   coefficient of thermal expansion, 168
   front flip, 161
   intermittent problems, 170
   non-ZIF, 161, 163
   slider, 161–162
   tabbed, 161, 163
contact cleaner spray, 25–26
continuity mode, digital multimeter, 90
corrosion damage, 165
counterfeit parts, 97
cracks in components, 166
crimp connectors, 36–38
   color code, 37
   crimper, 46–47
   soldering, 37
   wire gauge, 37
crimper, 46–47
current blasting, 182
current measurement, using digital multimeter, 90–91
cutting cases apart, 156–158
cutting tools:
   hot knife tip, 156–158
   razor saw, 156–157
   utility saw, 156–157
cyanoacrylate (*See* superglue)

darkened areas on PC board, 167
datasheet, integrated circuit, 177–178
DC power supply (*See* power supply)

DC voltage, 22–23
deformed components, 165
dental pick, 46
desoldering, 72–77
    add solder, 74–75
    clip instead of, 76–77
    flux pen, 74–75
    pump, 74–75
    solder sucker, 74–75
    solder wick, 73
    surface-mount parts, 84–86
    use a pin vise, 75–76
digital calipers, 55–56
digital multimeter:
    checking alkaline batteries, 87–88
    continuity mode, 90
    current measurement, 90–91
    diode test, 91–92
    features, 50–51
    general measurement procedure, 87–88
    piercing probe tip substitute, 89–90
    uses, 50–51
diode bridge, 181
diode test, digital multimeter, 91–92
diodes:
    checking, 117
    identification, 117–118
    in-circuit testing, 171
    rectifier, 118
    Schottky, 118
    signal, 118
    uses, 12–13
    varactors, 119
    Zener, 119
DIP (*See* integrated circuits, packages)

direct current (*See* DC voltage)
disassembly (*See* taking it apart)
divide and conquer, 179–180
dummy loads, 56–58

Edison, Thomas, 23
egg cartons, for screws and parts, 158–159
electrical tape, 30
electricity:
    AC voltage, 22–23
    DC voltage, 22–23
    Ohm's Law, 20
    power, 20
    resistance, 19
    units, 19
    water analogy, 19
electrolytic capacitors, 106–111
    axial leads, 106–107
    bulging, 164
    discharging, 5–6
    discharging, 108–110
    identification, 106
    in-circuit testing, 171
    in-circuit testing, 175–176
    leaking, 164
    likely failures, 167
    name brands, 111
    nonpolarized, 107
    radial leads, 106–107
    reliability, 111
    substitution, 110–111
    temperature rating, 111
    testing, 107
electronic waste disposal, 8
    rechargeable batteries, 210–211
    (*See also* RoHS)
electrostatic discharge (*See* ESD)
epoxy, 33–34
epoxy putty, 33–34

Index

ESD:
- antistatic desoldering pump, 74
- antistatic mat and wrist strap, 53–54
- input/output components, 168
- tweezers, safe, 45

ESR meter, 60–61
- for finding shorts, 182

e-waste disposal (*See* electronic waste disposal)

eye protection, 8

factory reset, 153–154
failure suspects, likely, 167–170
- intermittent problems, 170

FFC connectors, 160–163
film capacitors, 113
flat-screen TVs, 191–202
- backlights, 192–193
- blinking power LED, 200
- cracked screen, 201
- diagnosing problems, 196–202
- entire picture looks bad for all inputs, 201
- flashlight test for screen, 197
- main board, 193–194
- multiple columns of stuck-on or stuck-off pixels, 202
- no picture but backlight working, 201
- no picture, no backlight, 201
- no signs of power, 198
- only portions of picture look bad, 202
- panel assembly, 192–193
- picture okay but audio problems, 202
- power supply board, 193–194
- preliminary diagnosis, 197
- reset, 196–197
- Samsung open jig, 195
- secondary diagnosis, 197–199
- single row of lit or unlit pixels, 202
- software updating, 197
- symptoms and likely suspects, 199
- taking it apart, 194–196
- T-con board, 193–194
- turns off or on by itself, 200
- what's inside, 192–194
- wi-fi problems, 202
- won't turn on, 198–200

flux:
- acid-core, 69
- pen, 74–75
- removal, 73
- rosin-core, 69

flux pen, 74–75
foam surround, woofers, 221–228
freeze mist, 27–28
front flip connectors, 161
fs, 220–221
function generator, 64
fuses, 130–132
- failure, 131
- identification, 130–131
- pigtail, 131–132
- replacement, 131–132
- testing, 130–131
- thermal (*See* thermistors)

glues, 33–34
grommet (*See* strain-relief)

hairline cracks in components, 166
hand tools, 38–48
- crimper, 46–47
- dental pick, 46

hemostats, 45
JIS screwdrivers, 42
needle-nose pliers, 38
nut-driver set, 47–48
pin vise, 46
precision screwdrivers, 40–41
razor saw, 156–157
Samsung open jig, 195
security bits, 44
strain-relief pliers, 47
tweezers, 45
utility saw, 156–157
wire cutters, 38–39
wire stripper, 39–40
X-Acto knife, 40–41
heat gun, 54–55
heat shrink tape, 28–30
heat shrink tubing, 28–29
heat sink compound, 30, 80–81
   versus thermal pads, 82–83
heat sinks, 80–83
heat-sink grease (*See* heat sink
   compound)
heat-sinked devices, 80–83
   heat sink compound, 80–81
   mica insulators, 83
   soldering, 83
   thermal pads, 81–82
hemostats, 45
hidden screws, 154
hidden tabs, 155–156
high voltage, safety, 5–7
high-power components, 167

ICs (*See* integrated circuits)
iFixit, 42, 53, 54, 156, 177, 178,
   191, 202, 233
in-circuit component testing,
   170–176
   diodes, 171

   electrolytic capacitors, 171
   electrolytic capacitors,
      175–176
   joy of sets, 173, 176
   lift one lead, 172
   potentiometers, 171
   power the circuit, 173
   resistors, 171
   transistors, 171
   transistors, 173–174
   Zener diodes, 173–175
inductors:
   checking, 115
   identification, 115–116
   replacement, 116
   uses, 11–12
inspection, visual, 164–167
insulators, mica, 83
integrated circuits:
   datasheet, 177–178
   identification, 125
   packages, 126–127
   pin numbering, 126–127
   substitution, 127
   uses, 14
   voltage regulators, 124–125
intermittent problems, 169–170
IPA (*See* isopropyl alcohol)
isolation transformer, 64–65
   choosing, 65
   use, 64–65
isopropyl alcohol, 32–34
   flux removal, 72
   heat sink compound removal,
      81

Japanese Industrial Standard
   (*See* JIS)
jewelers screwdrivers (*See*
   precision screwdrivers)

JIS:
   screwdrivers, 42
   screws, 43–44
joy of sets, 173, 176

laptop computers, 202–206
   cracked screen replacement, 203–204
   finding replacement parts, 203
   keyboard replacement, 205–206
   replacing individual keys, 206
lasers, 7
leaking electrolytic capacitors, 164
LED TVs, 193
lightning damage, 152
Li-ion batteries (*See* rechargeable batteries)
likely failure suspects, 167–170
   intermittent problems, 170
liquid damage, 152, 165
lithium-ion batteries (*See* rechargeable batteries)
log, repair, 187–190
loudspeakers, 217–228
   blown tweeter, 217–218
   driver damage, 217–218
   driver replacement, 218–219
   listening test and fine-tuning, 219–220
   tweeter pads, 219–220
   two-way, three-way, 217
   woofer anatomy, 221–222
   woofer substitution, 220–221
   woofer surround repair, 221–228

magnetic screwdriver, 154–155
magnets, for clip leads, 36
magnifiers:
   headband, 51–52
   lamp, 51–52
melted insulation on wires, 166
meter:
   analog (*See* analog meters)
   ESR (*See* ESR meter)
   (*See also* digital multimeter)
mica insulators, 83
micrometer (*See* digital calipers)
microwave oven capacitors, 110
MOSFET amplifiers, troubleshooting, 183
MOSFETs, testing, 122–123
MOVs (*See* varistors)
multicolored tape, for labeling screw types, 158–159
multimeter (*See* digital multimeter)

nail polish, 31
needle-nose pliers, 38
NiCd batteries (*See* rechargeable batteries)
nickel-cadmium batteries (*See* rechargeable batteries)
nickel-metal hydride (*See* rechargeable batteries)
NiMH batteries (*See* rechargeable batteries)
NPN transistors (*See* bipolar transistors)
NTC (*See* thermistors)
nut-driver set, 47–48

ohmmeter (*See* digital multimeter)
Ohm's Law, 20
OLED displays, 193
OptiVISOR, 51–52

oscillator (*See* function generator)
oscilloscope, 62–63
   choosing, 62
   PC, 63
   probe compensation, 94–96
   probes, 93–94
   triggering, 96
   uses, 62
   using, 93–96

Panavise, 56–57
parallel connection, 20–22
parallel-cap technique, 175–176
PC board:
   coefficient of thermal expansion, 168
   trace repair, 78–79
PC oscilloscope, 63
PCB connectors, 160–163
Phillips, Henry F., 43
piercing probe tip substitute, 89–90
pin vise, 46, 75–76
PNP transistors (*See* bipolar transistors)
pole and throw terminology, 133–134
polyester capacitors (*See* film capacitors)
polypropylene capacitors (*See* film capacitors)
potentiometers:
   identification, 103
   in-circuit testing, 171
   intermittent problems, 170
   substitution, 105
   taper, 105
   testing, 103–104
   uses, 9–10
pots (*See* potentiometers)

power:
   distribution, 23
   formula for, 20
power adapters (*See* AC adapters)
power amps (*See* audio receivers and power amps)
power bricks (*See* AC adapters)
power supply, 59–60
   choosing, 59
   circuit, 181–182
   current blasting, 182
   troubleshooting, 180–182
   uses, 59
power surges, 138
precision screwdrivers, 40–41
probe compensation, for oscilloscope, 94–96
probes, oscilloscope, 93–94
pry tools, 155–156
PTC (*See* thermistors)

QFP (*See* surface-mount parts)
Qts, 220–221
quantum-dot displays, 193

radial leads, electrolytic capacitors, 106–107
receivers (*See* audio receivers and power amps)
rechargeable batteries, 207–209
   battery chemistry, 207–208
   capacity, 208
   dimensions, 208–209
   proper disposal, 210–211
   sub C, 208–209
rechargeable battery packs, 207–211
   identifying cell types, 207–209
   rebuilding, 209–210
   replace or repair, 207

rectifier diodes, 118
recycling (*See* electronic waste disposal)
refoaming woofers, 221–228
   resizing a surround, 225–226
   shimless surround repair, 227–228
   surround size and edge type, 224–226
   to shim or not to shim, 222–224
regulators (*See* voltage regulators)
relays, 139–141
   contact form, 140–141
   in audio receivers and power amps, 212–213
   intermittent problems, 170
   replacement, 140–141
   speaker protection, 213
   substitution, 140–141
   testing, 140
   uses, 15–16
remote controls, 228–231
   battery problems, 229
   camera test, 229
   keypad problems, 230
   replacement, 231
   user error, 228
repair, PC board trace, 78–79
repair forums, 177–178
repair log, 187–190
reset, flat-screen TVs, 196–197
reset, factory, 153–154
resistance, Ohm's Law, 20
resistors:
   color code, 99
   identification, 99–102
   in series and parallel, 21–22
   in-circuit testing, 171
   power rating, 102
   SMD marking codes, 100–101
   substitution, 102
   testing, 98
   uses, 9–10
right to repair movement, 178
RoHS (restriction of hazardous substances), 135–136

safety, 5–8
   childproofing, 8
   discharging electrolytic capacitors, 5–6
   electronic waste disposal, 8
   eye protection, 8
   high voltage, 5–7
   isolation transformer, 64–65
   lasers, 7
   solder and soldering, 7
   solvents and chemicals, 7
   working alone, 8
Samsung open jig, 195
schematic diagrams, 16–19
   component symbols, 18
   connections between wires, 18
   draw your own, 179
schematics (*See* schematic diagrams)
Schottky diodes, 118
scope (*See* oscilloscope)
screwdriver magnetizer, 154–155
screwdrivers:
   JIS, 42
   magnetic, 154–155
   precision, 40–41
screws:
   hidden, 154
   keeping track of which go where, 158
security bits, 44

series connection, 20–22
service manual, 177–178
signal diodes, 118
signal generator (*See* function generator)
sinewave generator (*See* function generator)
SMD (*See* surface-mount parts)
SMD Codebook, 118, 119, 125, 236
SOIC (*See* surface-mount parts)
solder, 68–70
    acid-core, 69
    eutectic, 70
    rosin-core, 69
    safety, 7
    Sn63Pb37, 68–70
    (*See also* soldering)
solder braid (*See* solder wick)
solder joints:
    bad, 166
    cold, 70
solder sucker, 74–75
solder wick, 73
soldering:
    cold solder joints, 70
    flux removal, 72
    heat-sinked devices, 83
    how to, 71–72
    safety, 7
    solder, 68–70
    surface-mount parts, 84–86
    temperature, 67–68
    tip size, 67–68
    (*See also* desoldering)
soldering iron, 48–50
    hot knife tip, 156–158
    tips, 49–50
soldering station (*See* soldering iron)

solderless connectors (*See* crimp connectors)
solvents, 32–33
    safety, 7
speaker protection relays, 213
speakers (*See* loudspeakers)
spill damage, 152
squarewave generator (*See* function generator)
static electricity (*See* ESD)
strain-relief pliers, 47
sub C rechargeable batteries, 208–209
substitution:
    AC adapters, 149
    analog meters, 143
    electrolytic capacitors, 110–111
    integrated circuits, 127
    potentiometers, 105
    relays, 140–141
    resistors, 102
    switches, 135–136
    transformers, 138–139
    transistors, 123–124
subwoofers (*See* woofers)
superglue, 33–34
surface-mount parts, 84–86
    capacitor marking code, 114
    desoldering, 84–86
    resistor marking codes, 100–101
    soldering, 84–86
switches, 132–136
    checking, 132
    intermittent problems, 170
    momentary, 135
    pole and throw terminology, 133–134
    power, 136
    replacement, 135–136

switches (*continued*)
   SPST, SPDT, DPDT, 133–134
   substitution, 135–136
symbols, component, 18

taking it apart, 154–163
   board connectors, 160–163
   cutting it apart, 156–158
   flat-screen TVs, 194–196
   hidden screws, 154
   hidden tabs, 155–156
   labeling screw types, 158–159
   pry tools, 155–156
   screws under feet and labels, 154
   take pictures, 160
tamperproof bits (*See* security bits)
tantalum capacitors, 111–112
   color code, 112
   likely failures, 167
tape, electrical, 30
tape, heat shrink, 28–30
T-con board, 193–194
Tesla, Nikola, 23
test equipment:
   digital multimeter, 50–51
   dummy load, 56–58
   ESR meter, 60–61
   function generator, 64
   isolation transformer, 64–65
   oscilloscope, 62–63
   power supply, 59–60
   soldering station, 48–50
testing:
   AC adapters, 145–146
   analog meters, 141
   bipolar transistors (NPN, PNP), 121–122
   components, 150

   electrolytic capacitors, 107
   fuses, 130–131
   MOSFETs, 122–123
   potentiometers, 103–104
   power transformers, 138
   relays, 140
   resistors, 98
   voltage regulators, 125
thermal compound (*See* heat sink compound)
thermal conductivity:
   heat sink compound, 82–83
   thermal pads, 82–83
thermal fuses (*See* thermistors)
thermal pads, 81–82
   versus heat sink compound, 82–83
thermal paste (*See* heat sink compound)
thermistors, 129–130
   identification, 130
   replacement, 130
Thiele/Small parameters, 220–221
third hand (*See* Panavise)
three-terminal regulators (*See* voltage regulators)
tie wraps (*See* wire ties)
tips, soldering iron, 49–50
TO-3, TO-92, TO-220 (*See* transistors, packages)
tools (*See* hand tools)
trace repair of PC board, 78–79
transformers, 137–139
   center-tapped, 137
   checking, 138
   failure, 138
   isolation, 64–65
   power, 137
   replacement, 138–139

signal, 137
substitution, 138–139
uses, 14–15
transistors:
    amplifier circuit, 186–187
    bipolar (NPN, PNP), 121
    failure modes, 120
    general purpose, 123–124
    identification, 119–120
    in-circuit testing, 171
    MOSFETs, 121
    packages, 120
    substitution, 123–124
    switching circuits, 183–185
    testing bipolar (NPN, PNP), 121–122
    testing MOSFETs, 122–123
    uses, 12–13
triggering, of oscilloscope, 96
troubleshooting, 176–187
    bipolar transistor amplifiers, 182–183
    divide and conquer, 179–180
    fix the cause, 180
    MOSFET amplifiers, 183
    power supplies, 180–182
    transistor circuitry, 182–187
    transistor switching circuits, 183–185
    understand the system, 177–179
    what is troubleshooting, 177
TVs, flat-screen (*See* flat-screen TVs)
tweeter pads, 219–220
tweeters, 217–218
tweezers, 45

USB oscilloscope, 63

vacuum tubes, 191
varactor diodes, 119
varistors, 128–129
    identification, 129
    replacement, 129
Vas, 220–221
vise, 56–57
visual inspection, 164–167
voltage regulators, 124–125
    checking, 125
voltmeter (*See* digital multimeter)

wall chargers (*See* AC adapters)
wall warts (*See* AC adapters)
water damage, 152
wire cutters, 38–39
wire gauge:
    crimp connectors, 37
    PC board trace repair, 78
wire stripper, 39–40
wire ties, 31–32
wires, melted insulation, 166
wire-to-board connectors, 160–163
woofers, 217–218
    anatomy, 221–222
    refoaming, 221–228
    shimless surround repair, 227–228
    substitution, 220–221
    surround repair, 221–228
    surround size and edge type, 224–226
working alone, 8

X-Acto knife, 40–41

Zener diodes, 119
zip ties (*See* wire ties)

About the Author

Mark Rumreich has worked for more than 25 years in the consumer electronics industry as a designer of analog and digital audio and video systems. He received a master's degree in electrical engineering from Purdue University and holds more than 60 US patents. He is the author of *The Car Stereo Cookbook: How to Design, Choose, and Install Car Stereo Systems*, now in its second edition.

Printed in Great Britain
by Amazon